COLLECTED WORKS OF
RICHARD J. CHORLEY

Volume 3

INTRODUCTION TO FLUVIAL PROCESSES

T0174765

INTRODUCTION TO
FLUVIAL PROCESSES

Edited by
RICHARD J. CHORLEY

Routledge
Taylor & Francis Group

LONDON AND NEW YORK

First published in 1969 by Methuen and Co. Ltd as part of 'Water, Earth and Man'
Published in this form in 1971

This edition first published in 2019
by Routledge
2 Park Square, Milton Park, Abingdon, Oxon OX14 4RN

and by Routledge
52 Vanderbilt Avenue, New York, NY 10017

Routledge is an imprint of the Taylor & Francis Group, an informa business

British Library Cataloguing in Publication Data
A catalogue record for this book is available from the British Library

ISBN: 978-0-367-22096-9 (Set)
ISBN: 978-0-429-27321-6 (Set) (ebk)
ISBN: 978-0-367-22106-5 (Volume 3) (hbk)
ISBN: 978-0-367-22179-9 (Volume 3) (pbk)
ISBN: 978-0-429-27331-5 (Volume 3) (ebk)

Publisher's Note
The publisher has gone to great lengths to ensure the quality of this reprint but
points out that some imperfections in the original copies may be apparent.

Disclaimer
The publisher has made every effort to trace copyright holders and would welcome
correspondence from those they have been unable to trace.

Introduction to
FLUVIAL PROCESSES

EDITED BY

Richard J Chorley

CONTRIBUTORS
M. A. Carson, R. J. Chorley, G. H. Dury, I. S. Evans,
R. W. Kates, B. A. Kennedy, M. J. Kirkby, S. A. Schumm
D. B. Simons, D. R. Stoddart, and P. W. Williams

METHUEN & CO LTD

First published in 1969
First published as a University Paperback in 1971
Reprinted twice
Reprinted 1977

© *1969 Methuen & Co Ltd*

Printed in Great Britain by
Richard Clay (The Chaucer Press), Ltd
Bungay, Suffolk

ISBN 0 416 68820 9

Distributed in the USA by
HARPER & ROW PUBLISHERS, INC.
BARNES & NOBLE IMPORT DIVISION

Contents

Preface to the Paperback Edition

This paperback originally formed part of a larger, composite volume entitled *Water, Earth, and Man* (Methuen and Co Ltd, London, 1969, 588 pp.), the purpose of which was to provide a synthesis of hydrology, geomorphology, and socio-economic geography. The present book is one of a series of three paperbacks, published simultaneously, which set out these themes separately under the respective titles:

Introduction to Physical Hydrology
Introduction to Fluvial Processes
Introduction to Geographical Hydrology.

The link with the parent volume is maintained by the retention of the Introduction, which gives the rationale for associating the three themes. The aim of this paperback is primarily to make available in a cheap and handy form one of these systematic themes. In doing so, however, it is hoped that the book will provide a constant reminder of the advantages inherent in adopting a unified view of the earth and social sciences, and, in particular, that the study of water in the widest sense presents one of the most logical means of increasing our understanding of the interlocking physical and social environments.

Acknowledgements

The editor and contributors would like to thank the following editors, publishers and individuals for permission to reproduce figures and tables:

Editors

American Journal of Science for figs. 2.II.8 and 7.II.6; *Annals of the Association of American Geographers* for figs. 10.II.2 and 10.II.3; *Bulletin of the American Association of Petroleum Geologists* for fig. 2.II.5; *Bulletin of the Geological Society of America* for figs. 2.II.4, 2.II.11, 2.II.17, and 5.I.6(*a*); *Bulletin of the International Association of Scientific Hydrology* for figs. 4.II.3, 4.II.5, and 5.I.3; *Journal of Geology* for figs. 2.II.18 and 3.II.6; *Marine Geology* for fig. 1.III.8; *Professional Geographer* for fig. 10.II.1; *Transactions of the American Geophysical Union* for figs. 2.II.20 and 5.II.3.

Publishers

American Association of Petroleum Geologists for figs. 1.III.6 and 1.III.7 from *Recent Sediments, Northwest Gulf of Mexico* by F. P. Shepard *et al.*; W. H. Freeman and Co., San Francisco, for fig. 3.II.10 from *Fluvial Processes in Geomorphology* by L. B. Leopold, M. G. Wolman, and J. P. Miller; Heinemann, London, for fig. 3.II.1 from *Water, The Mirror of Science* by K. S. Davis and J. A. Day; McGraw-Hill Book Co., New York for figs. 2.II.3, 2.II.6, and 2.II.14 from *Handbook of Applied Hydrology* by Ven Te Chow (Ed.); Merrill Books, Inc., Columbus, for fig. 3.II.2 from *Oceanography* by M. G. Gross; Oliver and Boyd, Edinburgh for fig. 1.III.3 from *Principles of Lithogenesis* by N. M. Strakhov; Prentice-Hall, Inc., New York, for figs. 3.II.3 and 3.II.9 from *Physical Geology* by L. D. Leet and S. Judson; Presses Universitaires de France for figs. 1.III.1 and 1.III.2 from *Climat et Erosion* by F. Fournier; Princeton University Press for figs. 11.II.2 and 11.II.3 from *The Quaternary of the United States* by H. E. Wright and D. G. Frey (Eds.); John Wiley and Sons, Inc., New York, for fig. 1.III.9 from *The Sea* (Vol. 3) by M. N. Hill (Ed.), fig. 2.II.13 from *Geohydrology* by R. J. M. DeWiest.

Individuals

The Director, Department of Lands and Forests, British Columbia, for the base maps for fig. 8.II(i).2; The Director, Geographical Branch, Department of Mines and Technical Surveys, Ottawa, for figs. 8.II(ii).2 and 8.II(ii).3 from the *Memoir* by J. R. Mackay; The Director, Geographical Branch, Office of Naval

Research, Washington, for fig. 2.II.16 from the *Technical Report* by A. Broscoe and figs. 2.II.21 and 5.II.4 from the *Technical Report* by M. A. Melton; The Director, Military Engineering Experiment Establishment, Christchurch, Hampshire, for fig. 2.II.1; The Director, Tennessee Valley Authority, Office of Tributary Development, Knoxville, Tennessee, for fig. 5.I.5 from the *Research Paper* by R. P. Betson; The Director, U.S. Army Engineer Experiment Station, Vicksburg, Mississippi, for fig. 2.II.1; The Director, U.S. Geological Survey for figs. 2.II.12, 2.II.21, 3.II.7, and for the use of the originals from which a number of figures in Chapters 7.II and 9.II were prepared.

Finally, the following thanks are also due:

Mr M. A. Church for data on Banks Island slope angles employed in Chapter 8.II(ii); Professor J. N. Jennings, Department of Biogeography and Geomorphology, Australian National University for valuable comments on Chapter 6.II; Mr M. Young, Miss R. King, and Mr M. J. Ampleford of the Drawing Office, Department of Geography, Cambridge University, for drawing figures for Chapters 2.II and 3.II.

The Editor and Publishers would like to thank Mrs D. M. Beckinsale for her painstaking and authoritative preparation of the Index, which has contributed greatly to the value of this volume.

Introduction

R. J. CHORLEY and R. W. KATES

Department of Geography, Cambridge University and Graduate School of Geography, Clark University

Who would not choose to follow the sound of running waters? Its attraction for the normal man is of a natural sympathetic sort. For man is water's child, nine-tenths of our body consists of it, and at a certain stage the foetus possesses gills. For my part I freely admit that the sight of water in whatever form or shape is my most lively and immediate form of natural enjoyment: yes, I would even say that only in contemplation of it do I achieve true self-forgetfulness and feel my own limited individuality merge into the universal.

(Thomas Mann: *Man and his Dog*)

1. 'Physical' and 'human' geography

Perhaps it is of the nature of scholarship that all scholars should think themselves to be living at a time of intellectual revolution. Judged on the basis of the references which they have cited (Stoddart, 1967, pp. 12–13), geographers have long had the impression that they were the immediate heirs of a surge of worthwhile and quotable research. There is good reason to suppose, however, that geography has just passed through a major revolution (Burton, 1963), one of the features of which has been profoundly to affect the traditional relationships between 'physical' and 'human' geography.

Ever since the end of the Second World War drastic changes have been going on in those disciplines which compose physical geography. This has been especially apparent in geomorphology (Chorley, 1965a), where these changes have had the general effect of focusing attention on the relationships between process and form, as distinct from the development of landforms through time. In the early 1950s geomorphologists, especially in Britain, were able to look patronizingly at the social and economic branches of geography and dismiss them as non-scientific, poorly organized, slowly developing, starved of research facilities, dealing with subject matter not amenable to precise statement, and denied the powerful tool of experimentation (Wooldridge and East, 1951, pp. 39–40). It is true that by this time most geographers had long rejected the dictum that physical geography 'controlled' human geography, but most orthodox practitioners at least paid lip service to the idea that there was a physical *basis* to the subject. This view was retained even though traditional geomorphology had little or nothing to contribute to the increasingly urban and industrial preoccupations of human geographers (Chorley, 1965b, p. 35), and its place in the subject as a whole was maintained either as a conditioned reflex or as increasingly embarrassing grafts on to new geographical shoots. American geographers, who had largely abandoned geomorphology to the geologists even before the war, tended to look more to climatology for their physical basis. However, despite the important researches of Thornthwaite and of more recent work exemplified by that of Curry [1952] and Hewes [1965], the proportion of articles relating to weather and climate appearing in major American geographical journals fell more or less steadily from some 37% in 1916 to less than 5% in 1967 (Sewell, Kates, and Phillips, 1968). Even in the middle of the last decade Leighly (1955, p. 317) was drawing attention to the paradox that instructors in physical geography might be required to teach material quite unrelated to their normal objects of research.

The problems of the relationships between physical and human geography facing Leighly were small, however, compared with those which confront us today. Little more than a decade has been sufficient to transform the leading edge of human geography into a 'scientific subject', equipped with all the quantitative and statistical tools the possession of which had previously given some physical geographers such feelings of superiority. Today human geography is not directed towards some unique areally-demarcated assemblage of information which can be viewed either as a mystical *gestalt* expressive of some 'regional personality' or simply as half-digested trivia, depending on one's viewpoint. In contrast, most of the more attractive current work in human geography is aimed at more limited and intellectually viable syntheses of the pattern of human activity over space possessing physical inhomogeneities, leading to the disentangling of universal generalizations from local 'noise' (Haggett, 1965). Today it is human geography which seems to be moving ahead faster, to have the more stimulating intellectual challenges, and to be directing the more imaginative quantitative techniques to their solution.

One immediate result of this revolution has been the demonstration, if this were further needed, that the whole of geomorphology and climatology is not coincident with physical geography, and that the professional aims of the former are quite distinct from those of the latter. This drawing apart of traditional physical and human geography has permitted their needs and distinctions, which had previously been obscure, to emerge more clearly. Perhaps the distinctions may have become too stark, as evidenced by current geographical preoccupations with a rootless regional science and with socio-economic games played out on featureless plains or within the urban sprawl. Perhaps this is what the future holds for geography, but it is clear that without some dialogue between man and the physical environment within a spatial context geography will cease to exist as a discipline.

There is no doubt that the major branches of what was previously called physical geography can exist, and in some cases already are existing, under the umbrella of the earth sciences, quite happily outside geography, and that they are probably the better for it. It is also possible that this will be better for geography in the long run, despite the relevance to it of many of the data and certain of the techniques and philosophical attitudes of the earth sciences. In their place a more meaningful and relevant physical geography may emerge as the product of a new generation of physical geographers who are willing and able to face up to the contemporary needs of the whole subject, and who are prepared to concentrate on the areas of physical reality which are especially relevant to the modern man-oriented geography. It is in the extinction of the traditional division between physical and human geography that new types of collaborative synthesis can arise. Such collaborations will undoubtedly come about in a number of ways, the existence of some of which is already a reality. One way is to take a philosophical attitude implied by an integrated body of techniques or models (commonly spatially oriented) and demonstrate their analogous application to both human and physical phenomena (Woldenberg and Berry, 1967; Haggett and Chorley, In press). Another way is to assume that the stuff of the physical world with which geographers are concerned are its resources – resources in the widest sense; not just coal and iron, but water, ease of movement, and even available space itself. In one sense the present volume represents both these approaches to integration by its concentration on the physical resource of water in all its spatial and temporal inequalities of occurrence, and by its conceptualization of the many systems subsumed under the hydrological cycle (Kates, 1967). In the development of water as a focus of geographical interest the evolution of a human-oriented physical geography and an environmentally sensitive human geography closely related to resource management is well under way.

2. Water as a focus of geographical interest

Water, Earth, and Man, both in organization and content, reflects the foregoing attitudes by illustrating the advantages inherent in adopting a unified view of the earth and social sciences. The theme of this book is that the study of water provides a logical link between an understanding of physical and social environments. Each chapter develops this theme by proceeding from the many aspects of water occurrence to a deeper understanding of natural environments and their fusion with the activities of man in society. In this way water is viewed as a highly variable and mobile resource in the widest sense. Not only is it a commodity which is directly used by man but it is often the mainspring for extensive economic development, commonly an essential element in man's aesthetic experience, and always a major formative factor of the physical and biological environment which provides the stage for his activities. The reader of this volume is thus confronted by one of the great systems of the natural world, the hydrologic cycle, following water through its myriad paths and assessing its impact on earth and man. The hydrologic cycle is a great natural system, but it should become apparent that it is increasingly a technological and social system as well. It has been estimated that 10% of the national wealth of the United States is found in capital structures designed to alter the hydrologic cycle: to collect, divert, and store about a quarter of the available surface water, distribute it where needed, cleanse it, carry it away, and return it to the natural system. The technical structures are omnipresent: dams, reservoirs, aqueducts, canals, tanks, and sewers, and they become increasingly sophisticated in the form of reclamation plants, cooling towers, or nuclear desalinization plants. The social and political system is also pervasive and equally complex, when one reflects on the number of major decision makers involved in the allocation and use of the water resources. White has estimated that for the United States the major decision makers involved in the allocation and use of water include at least 3,700,000 farmers, and the managers of 8,700 irrigation districts, 8,400 drainage districts, 1,600 hydroelectric power plants, 18,100 municipal water-supply systems, 7,700 industrial water-supply systems, 11,400 municipal sewer systems, and 6,600 industrial-waste disposal systems.

This coming together of natural potential and of human need and aspiration provides a unique focus for geographic study. In no other major area of geographic concern has there been such a coalescence of physical and human geography, nor has there developed a dialogue comparable to that which exists between geographers and the many disciplines interested in water. How these events developed is somewhat speculative. First, there is the hydrologic cycle itself, a natural manifestation of great pervasiveness, power, and beauty, that transcends man's territorial and intellectual boundaries. Equally important is that in the human use of water there is clear acknowledgement of man's dependence on environment. This theme, developed by many great teachers and scholars, (e.g. Ackerman, Barrows, Brunhes, Davis, Gilbert, Lewis, Lvovich, Marts, Powell, Thornthwaite, Tricart, and White), is still an important geographic concern, despite the counter trends previously described. Finally, there is no gainsaying the universal appeal of water itself, arising partly from necessity, but also from myth, symbol, and even primitive instinct.

The emergence of water as a field of study has been paralleled in other fields. In the application of this knowledge to water-resource development, a growing consensus emerges as to what constitutes a proper assessment of such development: the estimation of physical potential, the determination of technical and economic feasibility, and the evaluation of social desirability. For each of these there exists a body of standard techniques, new methods of analysis still undergoing development, and a roster of difficult and unsolved problems. Geographers have made varying contributions to these questions, and White reviewed them in 1963. Five years later, what appear to be the major geographic concerns in each area?

Under the heading of resource estimates, White cites two types of estimates of physical potential with particular geographic significance. The first is 'the generalized knowledge of distributions of major resources . . . directly relevant to engineering or social design'. While specific detailed work, he suggests, may be in the province of the pedologist, geologist, or hydrologist, there is urgent need for integrative measures of land and water potential capable of being applied broadly over large areas. The need for such measures has not diminished, but rather would seem enhanced by developments in aerial and satellite reconnaissance that provide new tools of observation, and by the widespread use of computers that provide new capability for data storage and analysis. In the developing world the need is for low-cost appraisal specific to region or project.

A second sort of estimate of potential that calls upon the skills of both the physical and human geographer is to illuminate what White calls 'the problem of the contrast between perception of environment by scientists . . . (and) others who make practical decisions in managing resources of land and water'. These studies of environmental perception have grown rapidly in number, method, and content. They suggest generally that the ways in which water and land resources receive technical appraisal rarely coincide with the appraisals of resource users. This contrast in perception is reflected in turn by the divergence between the planners' or technicians' expectation for development and the actual course of development. There are many concrete examples: the increase in flood damages despite flood-control investment, the almost universal lag in the use of available irrigation water, the widespread rejection of methods of soil conservation and erosion control, and the waves of invasion and retreat into the margins of the arid lands. Thus a geography that seeks to characterize environment as its inhabitants see it provides valued insight for the understanding of resource use.

In 1963 White differentiated between studies of the technology of water management and studies of economic efficiency. Today one can suggest that, increasingly, technical and economic feasibility are seen as related questions. The distinction between these areas, one seen as the province of the engineer and hydrologist, the other as belonging to the economist and economic geographer, is disappearing, encouraged by the impressive results of programmes of collaborative teaching and research between engineering and economics (e.g. at Stanford and Harvard Universities). In this view, the choice of technology and of scale is seen as a problem of cost. The choice of dam site, construction material, and height depends on a comparison of the incremental costs and of the incremental benefits arising from a range of sites, materials, and heights. This decision can be simultaneously related through systems analysis to the potential outputs of the water-resource system.

The methodology for making such determinations has probably outrun our understandings of the actual relationships. The costs and benefits of certain technologies are not always apparent, nor are all the technologies yet known. Geographic research on a broadened range of resource use and specific inquiry into the spatial and ecological linkages (with ensuing costs) of various technologies appears to be required. Indeed, as the new technologies of weather forecasting and modification, desalinization, and cross-basin transport of water and power expand, the need for such study takes on a special urgency.

Finally, there appears to be a growing recognition that much of what may be socially important in assessing the desirability of water-resource development will escape our present techniques of feasibility analysis for much time to come. The need for a wider basis of choice to account for the social desirability of water-resource development persists and deepens as the number of water-related values increase and the means for achieving them multiply. A framework for assessing social desirability still needs devising, but it could be hastened by

careful assessment of what actually follows water-resource development. There
is much to be learned from the extensive developments planned or already
constructed. However, studies such as Wolman's [1967] attempt to measure the
impact of dam construction on downstream river morphology or the concerted
effort to assess the biological and social changes induced by the man-made lakes
in Africa are few and far between. Studies built on the tradition of geographic
field research but employing a rigorous research design over an extended period
of observation are required. Geographers, freed from the traditional distinction
between human and physical geography and with their special sensitivity towards water, earth, and man, have in these both opportunity and challenge.

REFERENCES

ACKERMAN, E. A. [1965], The general relation of technology change to efficiency in
water development and water management; In Burton I. and Kates, R., Editors,
Readings in Resource Management and Conservation (Chicago), pp. 450–67.
BURTON, I. [1963], The quantitative revolution and theoretical geography; *The
Canadian Geographer*, 7, 151–62.
BURTON, I. and KATES, R. [1964], The perception of natural hazards in resource
management; *Natural Resources Journal*, 3, 412–41.
CHORLEY, R. J. [1965a], The application of quantitative methods to geomorphology; In
Chorley, R. J. and Haggett, P., Editors, *Frontiers in Geographical Teaching*
(Methuen, London), pp. 147–63.
CHORLEY, R. J. [1965b], A re-evaluation of the geomorphic system of W. M. Davis;
In Chorley, R. J. and Haggett, P., Editors, *Frontiers in Geographical Teaching*
(Methuen, London), pp. 21–38.
CURRY, L. [1952], Climate and economic life: A new approach with examples from the
United States; *Geographical Review*, 42, 367–83.
HAGGETT, P. [1965], *Locational Analysis in Human Geography* (Arnold, London),
339 p.
HAGGETT, P. and CHORLEY, R. J. [1969], *Network Models in Geography* (Arnold,
London).
HEWES, L. [1965], Causes of wheat failure in the dry farming region, Central Great
Plains, 1939–57; *Economic Geography*, 41, 313–30.
HUFSCHMIDT, M. [1965], The methodology of water-resource system design; In
Burton, I. and Kates, R., Editors, *Readings in Resource Management and Con-
servation* (Chicago), pp. 558–70.
KATES, R. W. [1967], Links between Physical and Human geography; In *Introductory
Geography: Viewpoints and Themes* (Washington), pp. 23–31.
LEIGHLY, J. [1955], What has happened to physical geography?; *Annals of the
Association of American Geographers*, 45, 309–18.
SEWELL, W. R. D., Editor [1966], Human Dimensions of Weather Modification;
University of Chicago, Department of Geography, Research Paper 105, 423 p.
SEWELL, W. R. D., KATES, R. W., and PHILLIPS, L. E. [1968], Human response to
weather and climate; *Geographical Review*, 58, 262–80.
STODDART, D. R. [1967], Growth and structure of geography; *Transactions of the
Institute of British Geographers*, No. 41, 1–19.

WHITE, G. F. [1963], Contribution of geographical analysis to river basin development; *Geographical Journal*, **129**, 412–36.

WHITE, G. F. [1968], *Strategies of American Water Management* (Ann Arbor).

WOLDENBERG, M. J. and BERRY, B. J. L. [1967], Rivers and central places: Analogous systems?; *Journal of Regional Science*, **7** (2), 129–39.

WOLMAN, M. G. [1967], Two problems involving river channel changes and background observations; In Garrison, W. L. and Marble, D. F., Editors, *Quantitative Geography: Part II Physical and Cartographic Topics* (Northwestern University), pp. 67–107.

WOOLDRIDGE, S. W. and EAST, W. G. [1951], *The Spirit and Purpose of Geography* (Hutchinson, London), 176 p.

1.III. World Erosion and Sedimentation

D. R. STODDART

Department of Geography, Cambridge University

The mean elevation of the continents is $+840$ m and that of the oceans $-3,800$ m. Extreme elevations are, respectively, $+8,882$ m and $-11,500$ m. The mean elevation of the earth's surface, with respect to present sea-level, is $-2,440$ m. Erosion processes are constantly transferring material from the lands to the sea in the way that James Hutton described in 1795 in his *Theory of the Earth*. This section considers the magnitudes and areal variations of both erosion and sedimentation on a world scale.

1. Contemporary erosion: magnitude and patterns

Gross rates of denudation of the land surface are given by measuring the sediment load of rivers at their mouths. In practice, because of the difficulty of measuring bed load, rates are based on suspended sediment load and in some cases also solution load. Compilation of data from rivers throughout the world can thus be used to identify patterns of present denudation, especially when reduced to some common unit, such as metres per thousand years ($m/10^3$ yr). Because of the multivariate controls of rate of erosion, including relief, lithology, climate, and human use, these studies can yield only a first-order approximation, and it is not surprising that published syntheses are often mutually inconsistent, especially when based on sediment yields from basins of widely differing size.

Corbel studied total erosion for different temperature zones, in terms of three humidity and two relief categories. He found that erosion rates vary inversely with temperature, being lowest in the tropics; within each temperature zone they vary directly with humidity. Erosion is in all cases greater in mountainous areas than in plains, but the disparity between the two relief types is least in the tropics (a factor of 2), greatest in the temperate regions (a factor of 4–5), and intermediate in cold regions (a factor of 3–4). Figures for erosion in each category can be multiplied by area to give total erosion, and then averaged to give a mean world figure for unglaciated lands of $28·3$ $m^3/km^2/yr$, or roughly $0·03$ m/ 10^3 yr. This agrees remarkably with Dole and Stabler's estimate in 1909 of a rate of 1 ft in 9,000 yr ($0·034$ $m/10^3$ yr) for rivers in the continental United States. According to Corbel, all tropical areas, except for mountainous humid areas, are being eroded more slowly than this, and the temperate and cold lands, except for arid temperate plains, more rapidly. This pattern would presumably reflect more rapid mechanical weathering in cold lands, and would discount the effect of rapid chemical weathering and high rainfall in the tropics.

Other studies on a similar scale, using comparable data, yield rather different conclusions. Fournier studied suspended sediment yield in seventy-eight basins ranging in size from 0·0025 to 1·06 × 10⁶ km², correlating yield with a climatic parameter p^2/P, where p is the rainfall of the month with greatest rainfall and P the mean annual rainfall. The scatter of basins in terms of sediment yield and p^2/P, while generally confirming the expected increase of sediment yield with increase in rainfall, showed a grouping in terms of relief, into: (*a*) basins with

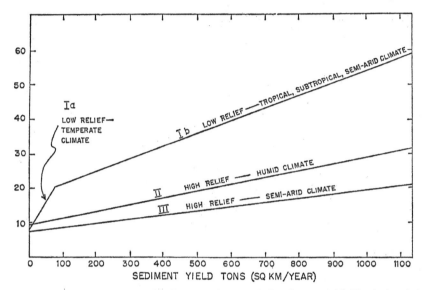

Fig. 1.III.1 Relationship between climate and suspended sediment yield. Vertical scale is p²/P (After Fournier, 1960).

low relief; (*b*) basins with high relief and a humid climate; and (*c*) basins with high relief and a semi-arid climate. It is thus impossible to relate rainfall to erosion without taking relief into account. Fournier derives a general empirical equation which fits his data, and which can be used for predicting sediment yield when climate and relief are known:

$$\log E \text{ (tons/km}^2\text{/yr)} = 2\cdot65 \log (p^2/P)\text{(mm)} + 0\cdot46 \log \bar{H} . \tan \phi - 1\cdot56$$

where E is suspended sediment yield, $\bar{H}$ mean height, and ϕ mean slope in a drainage basin. If $\bar{H} . \tan \phi$ is not readily available, basins may be grouped into one of the three classes already distinguished, and erosion calculated from the regression equations for each group (fig. 1.III.1): using this approximate method, Fournier has mapped the world distribution of erosion based on suspended sediment yield (fig. 1.III.2). He finds maximum rates in the seasonally

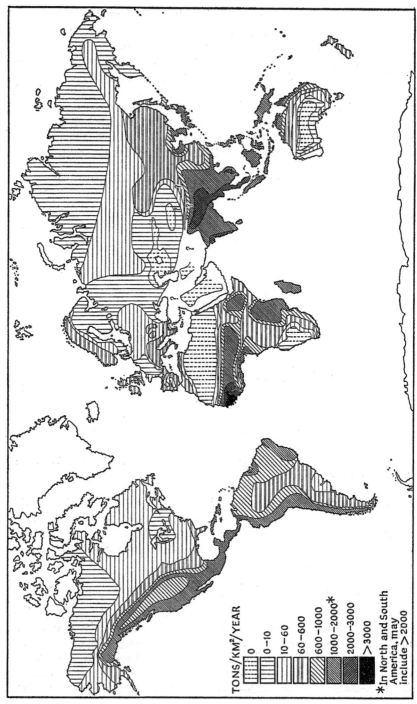

TONS/KM²/YEAR

0
0–10
10–60
60–600
600–1000
1000–2000*
2000–3000
>3000

*In North and South
America, may
include >2000

Fig. 1.III.2 World distribution of erosion (After Fournier, 1960).

TABLE I.III.I Rates of erosion of the continents

Continent	Area (km² × 10⁶)	Suspended sediment yield (tons × 10⁶)	Dissolved load yield (tons × 10⁶)	Denudation rate (tons/km²)	
				Mechanical	Chemical
Europe	9·67	420	305	43·0	32·0
Asia	44·89	7,445	1,916	166·0	42·0
Africa	29·81	1,395	757	47·0	25·2
North and Central America	20·44	1,503	809	73·0	40·0
South America	17·98	1,676	993	93·0	55·0
Australia	7·96	257	88	32·1	11·3

Source: Strakhov [1967]

humid tropics, declining in the equatorial regions where the seasonal effect is lacking and in the arid regions where the total amount of runoff is low. In the deserts long-distance transport of sediment, except by wind, is nil. The rate of erosion rises again in the seasonally wet Mediterranean lands, but over the temperate and cold regions it is low except in mountainous areas. While both Fournier and Corbel agree on the importance of climatic controls on erosion rates, in practice they invert the trend.

Fournier's conclusions are generally supported by Strakhov, using similar data on sediment yield in rivers. Strakhov's sample of rivers vary in drainage area from 13·4 to 7,050 × 10^3 km², in discharge from 11 to 3,187·5 km³/yr, and in sediment yield from 0·82 to 1,000 × 10^6 tons/yr. The tropical areas of intense chemical weathering are being eroded most rapidly, with sustained rates in the high-rainfall, high-relief areas of south-east Asia of 390 tons/km²/yr. Strakhov's rates (fig. 1.III.3 and Table 1.III.1) are in general less than Fournier's, sometimes by an order of magnitude; and this may reflect abnormally high rates, caused by human interference and accelerated soil erosion during historic time, in Fournier's sample. Thus Strakhov's rates may be geologically more 'normal' (Douglas, 1967).

Because of the problems of standardizing data from basins of differing magnitudes, it is possible to proceed by taking a series of small basins and extrapolating from them. This has been done in the western United States by Langbein and Schumm, using both stream-gauging and reservoir-fill data. Schumm found the following relationship between sediment yield and rainfall (standardized to a mean annual temperature of 50° F), sediment mass having been converted to erosion rates by assuming 1 ton of sediment to be equivalent to 4·34 × 10^{-7} ft/yr:

TABLE 1.III.2

Gauging station data		Reservoir-fill data	
Effective precipitation	m/10^3 yr	Effective precipitation	m/10^3 yr
10	0·088	8–9	0·186
10–15	0·104	10	0·155
15–20	0·073	11	0·198
20–30	0·073	14–25	0·149
30–40	0·052	25–30	0·189
40–60	0·030	30–38	0·104
		38–40	0·073
		40–55	0·064
		55–100	0·058

Source: Schumm [1963]

These data demonstrate a maximum sediment yield with a rainfall of 250–350 mm/yr: with a lower rainfall runoff is inadequate, and with a higher rainfall vegetation growth hinders the entrainment of sediment. Much higher rates than

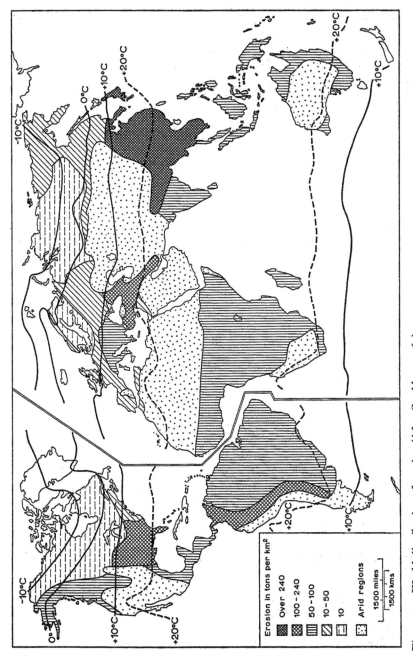

Fig. 1.III.3 World distribution of erosion (After Strakhov, 1967).

those listed in Table 1.III.2 are found in small basins in unconsolidated material (up to 12·8 m/10³ yr in loess badlands), and there is a general relationship between sediment yield and basin size, yield decreasing with the −0·15 power of basin area. Over the whole Mississippi basin, for example, erosion rate based on suspended load is only 0·04 m/10³ yr.

In addition to variation in sediment yield with rainfall, Schumm found sediment yield to be an exponential function of relief:

$$\log S \ (\text{acre-ft/sq. mile}) = 27\cdot35R - 1\cdot187$$

where R is the relief–length ratio. Similarly, denudation rates, standardized for a basin area of 3,885 km², are also a function of relief:

$$\log D \ (\text{ft/10}^3 \ \text{yr}) = 26\cdot866H - 1\cdot7238$$

where H is the relief–length ratio. Schumm's data indicate an average *maximum* rate of denudation of 0·9 m/10³ yr, and it is also clear that, though sediment

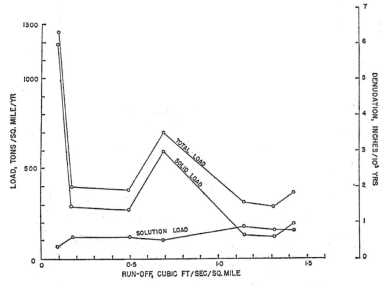

Fig. 1.III.4 Relationship between erosion rates and runoff (After Judson and Ritter, 1964).

yield is not a linear function of rainfall, meaningful distinctions can be made between arid, semi-arid, and humid areas. It is not clear from these data what happens in the hot wet lands of the tropics: Fournier's work would suggest that at rainfalls greater than 1,500 mm sediment yield rises to a new peak, and a similar inference can be drawn from Judson and Ritter's survey of regional rates of erosion in the United States (fig. 1.III.4).

In view of the problems raised in this survey of the world patterning of erosion by rivers, it is clearly premature to attempt to estimate the net sediment input in the erosion–depositional system. Confining attention to river sediments, moreover, neglects dissolved load, and also other modes of erosion, particularly by wind and sea. Dissolved load, which Clarke estimated to be $1·5$ km^3/yr over the earth, is important both in the denudation of the lands and also in the supply of minerals to the sea, for direct precipitation and for conversion to plant and animal skeletons and subsequent accumulation. Thus silica is being supplied to the sea at the rate of 320×10^6 tons/yr and calcium carbonate at the rate of 560×10^6 tons/yr. Rates of chemical denudation published by Strakhov vary from $3·9$ to 290 tons/km^2, whereas those of mechanical denudation range from $3·9$ to $2,000$ tons/km^2. Mechanical denudation adjusts much more rapidly than chemical to changes in flow conditions, for above a threshold value chemical denudation rates are a function of the availability of solutes rather than of discharge. Hence, while in a given basin the concentration of suspended load in rivers increases with increasing discharge, that of dissolved load decreases. In the United States the maximum dissolved load of 125–150 tons/mile^2/yr is reached with a runoff of about 250 mm, i.e. with an effective precipitation greater than that giving maximum solid sediment yield. With increasing relief the proportion of dissolved load decreases because of the much greater yield of solid load.

Rates of coastal erosion are of interest, if only because of the prolonged debate in the nineteenth century over the relative efficiency of marine and sub-aerial denudation. Many coasts in soft rocks are retreating rapidly: in the Sea of Azov at up to 12 m/yr, on the east coast of England at $1·5$–5 (maximum 11) m/yr, on the Polish Baltic coast and on Cape Cod at 1 m/yr. Harder rocks retreat more slowly, however (Normandy chalk cliffs $0·3$ m/yr; tidal notches on coral limestone coasts $0·001$ m/yr), and many coasts are actively aggrading. Kuenen estimates that the world's shorelines, 370×10^3 km long, yield $0·12$ km^3/yr of sediment, only 1% of that yielded by fluvial erosion.

These estimates of fluvial and coastal erosion give a world total annual sediment loss to the lands of $13·6$ km^3, or, averaged over the surface of the globe, 27×10^{-9} m/yr. This figure can have little real meaning, both because of the zonal distribution of rainfall and runoff, and the azonal distribution of upland areas, both combining to give a complex mosaic of erosion rates which are only beginning to be understood on the basin level. In terms of the azonal controls, a fundamental distinction must be made between the shield areas of the globe, covering 105×10^6 km^2 with mean elevation $0·75$ km, and the fold mountain belts, with half the area and twice the mean height (42×10^6 km^2; $1·25$ km). It should be noted also that the proportions of different kinds of rock vary in these different provinces, often in ways not indicated by conventional geological mapping, and that even on a continental scale lithology provides a powerful control of erosion rates (contrast, for example, Andean America with the Brazilian shield; 82% of the suspended sediment at the mouth of the Amazon comes from the Andes, according to Gibbs [1967]).

2. Contemporary sedimentation

Sediment loads from the continents are being transferred to ocean basins of differing sizes and shapes, and which differ widely in the proportions of land surface run-off they receive. Lyman has calculated that 68·5% of the world land area drains into the Atlantic, 44% in the Arctic, 26% into the Indian, and only

TABLE I.III.3 Magnitudes of world's oceans

Ocean	Area, km² × 10⁶	Volume, km³ × 10⁶	Mean depth, m	% world ocean area
Pacific	166·24	696·19	4,188	
With adjacent seas	181·34	714·41	3,940	50·1
Atlantic	86·56	323·37	3,736	
With adjacent seas	94·31	337·21	3,575	26·0
Indian	73·43	284·35	3,872	
With adjacent seas	74·12	284·61	3,840	20·5
Arctic	9·48	12·61	1,330	
With adjacent seas	12·26	13·70	1,117	3·4
World ocean	362·03	1,349·93	3,729	

Source: Menard and Smith [1966]

11% into the largest ocean, the Pacific. As a result, deep-sea sedimentation rates are almost an order of magnitude greater in the Atlantic than in the Pacific, averaging in the Holocene 0·088 m/10³ yr in the former and 0·01 m/10³ yr in the latter. As a first-order generalization, an erosion rate of 0·03 m/10³ yr in the great river basins (i.e. not including the deserts and other areas of low erosion) may be compared with a mean deposition rate of 0·01 m/10³ yr over the water 70% of the surface of the globe. Table I.III.3 shows the area, volume, and depth of the world's oceans, and Table I.III.4 the proportions of each ocean in different physiographic provinces.

A. Deltas

The 14 km³ yielded every year by erosion of the continents is provided mainly by areal (slope) and linear (rivers, coasts) processes: its pattern of deposition is by contrast primarily punctiform. Figure I.III.5 shows the world's twenty largest drainage basins, all with an area greater than 9×10^5 km²: in these, erosion products from almost 30% (437×10^5 km²) of the land area of the earth are being delivered to the twenty deltas and estuaries mapped. Apart from dissolved load which is added to the oceans, an average of 60% of the solid load of rivers is deposited at their mouths. Figure I.III.5 also demonstrates the wide range of solid discharge of these major rivers, from 10 to $1,800 \times 10^6$ tons/yr: some rivers with smaller basins, notably the Mekong and the Irrawaddy, have much higher sediment loads than even a river like the Congo with three times their discharge. Holeman [1968] has listed the major rivers of the world in terms of

TABLE I.III.4 Physiographic provinces of world's oceans

Ocean	Continental shelf and slope	Continental rise	Ocean basin	Mid-ocean ridge	Trench	Other
Pacific	13·1	2·7	43·0	35·9	2·7	2·5
Atlantic	17·7	8·0	39·3	32·3	0·7	2·0
Indian	9·1	5·7	49·2	30·2	0·3	5·4
Arctic	68·2	20·8	0	4·2	0	6·8
World ocean	15·3	5·3	41·8	32·7	1·7	3·1

Figures as percentages.
Source: Menard and Smith [1966]

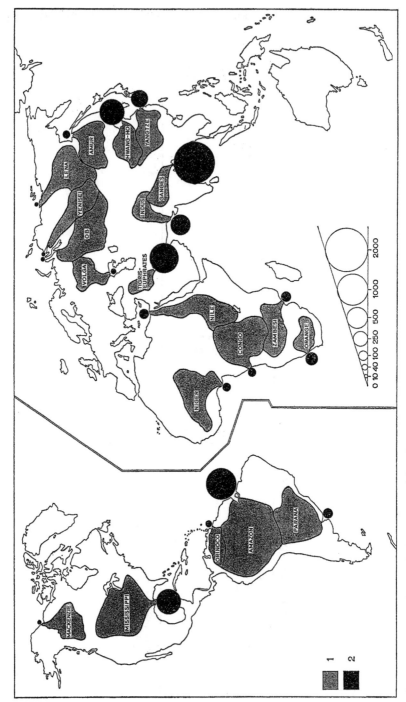

Fig. 1.III.5 World's largest drainage basins (1) and magnitude of solid load (2) (Data from Strakhov, 1967).

mean annual suspended sediment yield, rather than basin area (Table 1.III.5).
It is instructive to compare the data in Holeman's list with that mapped in Figure
1.III.5.

TABLE 1.III.5 Rivers of the world ranked by sediment yield

| River | Drainage basin 10^3 km^2 | Average annual suspended load | | Average discharge at mouth 10^3 cfs |
		Metric tons $\times 10^6$	Metric tons/km^2	
Yellow	673	1,887	2,804	53
Ganges	956	1,451	1,518	415
Brahmaputra	666	726	1,090	430
Yangtze	1,942	499	257	770
Indus	969	435	449	196
Ching	57	408	7,158	2
Amazon	5,776	363	63	6,400
Mississippi	3,222	312	97	630
Irrawaddy	430	299	695	479
Missouri	1,370	218	159	69
Lo	26	190	7,308	—
Kosi	62	172	2,774	64
Mekong	795	170	214	390
Colorado	637	135	212	5·5
Red	119	130	1,092	138
Nile	2,978	111	37	100

Source: Holeman [1968]

The Mississippi River, with a drainage basin area of 29×10^5 km^2, discharges
590 km^3 of water a year. The present sediment load is about 450×10^6 tons/yr,
containing 40% silt, 50% clay, and 2% sand. The late Pleistocene and Recent
delta covers 114×10^3 km^2, of which 28·5 are in the present deltaic plain, 45·3
on the continental shelf, 22·0 on the continental slope, and 18·1 in a submarine
bulge built during the last glacial low level of the sea. This late Quaternary delta
is estimated to contain 33,400 km^3, deposited at the rate of $1,570 \times 10^6$ tons/yr
when the river was entrenching and of 635×10^6 tons/yr when it was aggrading
as sea-level rose. The present distinctive birdfoot delta has been built entirely
since A.D. 1500, during the dumping of a further 110 km^3 of sediment (fig.
1.III.6).

These rates are so great that they cause easily measurable changes in topography round delta mouths, in striking contrast to the difficulty of measuring
erosion distributed over much greater areas. In the Mississippi Delta accretion
rates of more than 0·3 m/yr have been measured down to depths of 200 m where
the delta is building out on to the continental slope. Maximum rates on delta-
front platforms reach 0·3–0·45 m/yr over the last seventy years, falling away
from the delta to less than 0·03 m/yr. These are figures for wet sloppy sediment,
and should be multiplied by 0·5 to give a net accretion rate (see fig. 1.III.7). They
may be compared with rates of 0·12–0·3 m/yr $\times 10^3$ measured in the coastal

lagoons along much of the rest of the Gulf of Mexico coast, and are similar to rates measured for other great deltas of the world. Minor coastal accumulation features, such as salt marshes, have typical vertical growth rates of 0·01 m/yr, but are quantitatively of little significance in the planetary sediment budget.

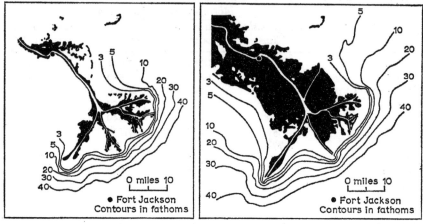

Fig. 1.III.6 Growth of the birdfoot delta of the Mississippi River 1874 (*left*)—1940 (*right*) (After Scruton, 1960).

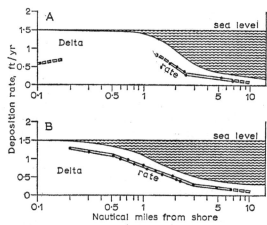

Fig. 1.III.7 Rates of accretion in the Mississippi Delta. A—Main Pass; B—North Pass (After Scruton, 1960).

B. Continental shelves and slopes

Continental shelves have a mean width of 75 km, mean edge depth of 130 m, and mean slope of 0° 07′; Kuenen calculates their area to be 30 × 10⁶ km² or 8% that of the oceans. Being adjacent to the lands, they collect most of the terrigenous solid sediments yielded by them; being relatively shallow, these deposits are subject to considerable reworking by marine action. Above all, the complexity of sedimentation on the shelves results from the fact that they lie within the range of Pleistocene eustatic sea-level shifts, which extended down to at least −150 m.

The sediment pattern on modern shelves consists therefore of deposits laid down when the shelves were emergent in the Würm, partly covered by transgressive marine deposits (transgressive from sea to land) and by terrigenous deltaic and coastal deposits of the present stillstand (transgressive from land to sea). Modern sedimentation is small on the outer parts of continental shelves, and is rapid on the inner parts and in nearshore basins. Hayes has studied the distribution of sediment types on the inner continental shelves in detail (Table 1.III.6). He finds that mud is most abundant off areas with high temperature

TABLE 1.III.6 Distribution of sediments on the inner continental shelf in terms of coastal climates

Coastal climatic zone	Known bottom sediments (%)						Unknown (%)
	Rocky	Gravel	Coral	Shell	Sand	Mud	
Rainy tropical	3·2	0·3	12·3	4·4	31·4	48·5	21·2
Sub-humid tropical	5·2	1·4	13·5	4·5	38·4	37·0	15·8
Warm semi-arid	3·7	—	8·6	4·2	59·5	24·0	8·5
Warm arid	4·4	—	7·3	4·8	52·1	31·4	24·2
Hyper-arid	9·1	0·6	20·9	12·0	44·5	12·8	33·6
Rainy subtropical	11·7	0·4	3·8	6·6	54·3	23·2	5·4
Summer-dry subtropical	26·1	4·1	2·1	2·7	37·3	27·7	11·9
Rainy marine	26·2	—	2·4	2·4	63·3	5·7	30·0
Wet-winter temperate	29·7	6·3	—	1·6	53·6	8·9	8·6
Rainy temperate	18·6	9·1	—	4·8	48·2	19·2	4·8
Cool semi-arid	—	—	—	7·1	92·9	—	—
Cool arid	20·2	4·8	2·4	3·6	52·8	16·3	20·8
Subpolar	30·8	14·9	—	3·5	39·3	11·5	7·0
Polar	20·8	16·2	—	4·8	43·1	15·1	9·2
Total	13·3	4·1	6·4	4·8	43·5	28·0	—

Source: Hayes [1967]

and high rainfall; sand is everywhere abundant, but with a maximum in areas of moderate temperature and rainfall and in all arid (except extremely cold) areas; gravel is most common off areas of low temperature; rock is most frequent in cold areas; coral is most abundant in areas with high temperature; and the distribution of shell is not related to climate. Figure 1.III.8, from Hayes, shows the climatic control in terms of precipitation and temperature of mud, sand, and gravel distribution; these diagrams are comparable to those used by Peltier in his delimitation of world morphoclimatic zones (see Chapter 10.II). In view of the limitations of the data, the known differences in shelf topography,

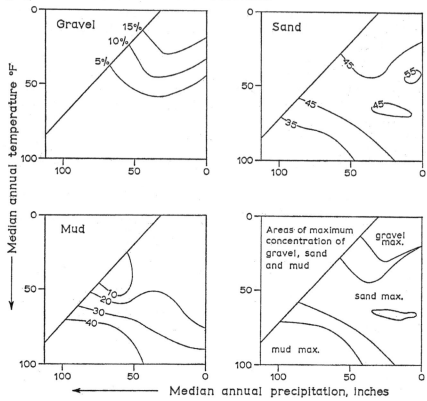

Fig. 1.III.8 Distribution of sediments on the inner continental shelves in terms of climatic controls (After Hayes, 1967).

and the importance of Pleistocene relict sediments, the correlation between sediment type and present coastal climate is remarkable.

Continental slopes have a mean width of 20–100 km, and form a transition zone between the shelf edge and the deep sea floor. Mean slope (contrary to many impressions) is only 4° 17′ for the first 1,800 m (2° 55′ in the Indian Ocean, 3° 05′ in the Atlantic, 5° 20′ in the Pacific), but the topography is fairly rugged because of the incision of submarine canyons in the slopes, cutting back into the shelves. Continental slopes received terrigenous sediment during Pleistocene low sea-levels: aggradation rates have been higher on the slopes than on the outer shelves. The slope is not simply an aggradation surface, however, comparable to deltaic foresets; it is too irregular, and Cretaceous and Tertiary sediments outcrop on it.

Submarine canyons are important as channels through which sediments from the lands are funnelled to the deep sea floor. Daly in 1936 first proposed that they were cut by turbidity currents, dense suspensions of sediment flowing down-slope, and Kuenen in 1950 experimentally demonstrated the reality of such

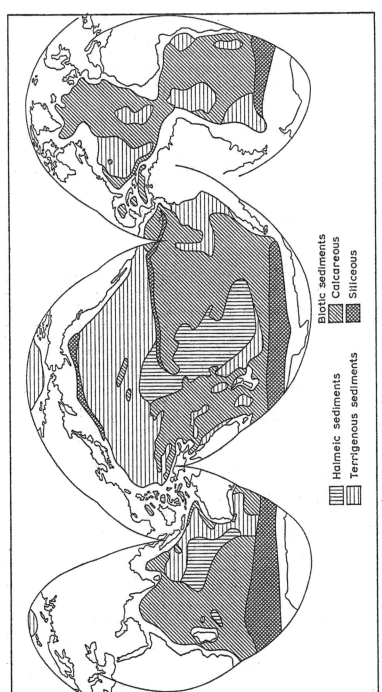

Fig. 1.III.9 Distribution of deep-sea sediments (After Arrhenius, 1963).

Biotic sediments
Calcareous
Siliceous

Halmeic sediments
Terrigenous sediments

currents. Whether turbidity currents can actually erode canyons remains open to doubt, but fine sands and other sediments are certainly transported by them through canyons and deposited in cones at their mouths. Large canyons have been identified off the mouths of many large rivers, including the Hudson and the Congo.

C. Deep ocean floor

Our knowledge of the deep ocean floor and its sediments began with the voyage of H.M.S. *Challenger* in 1872–6; recent work has demonstrated the diversity of deep-sea topography. The mid-ocean ridges occupy 32·7% of the oceans, basins proper 41·8%, and the continental rise at the foot of the continental slopes 5·3%; proportions differ between different oceans (Table 1.III.3). Sedimentation on the floor is of two main types: pelagic, formed by the slow settling of deposits far from land, and terrigenous.

(a) Pelagic deposition

Pelagic deposits consist of either biogenic material, diagenetic deposits, or clays. Table 1.III.7, from Kuenen, shows the proportions of different types in each

TABLE 1.III.7 Distribution of pelagic sediments

Sediment type	Area km² × 10³	% of deep ocean	% of total sea	Relative rate of deposition	% of total sediment volume
Calcareous sands and oozes	127·9	47·7	35·4	3	71·9
Brown (red) clay	102·2	38·1	28·3	1	19·2
Siliceous oozes	38·0	14·2	10·5	1·25	8·9
Total	268·1	100·0	74·2		100·0

Source: Kuenen [1950]

ocean, and Figure 1.III.9 their distribution. Calcareous oozes cover 128 × 10⁶ km² of floor: they are formed from the tests of foraminifera and other organisms settling to the ocean floor. At deeper levels calcium carbonate is dissolved by sea-water, and the 'compensation depth', below which calcareous deposits are rare, is 4,500–5,000 m. Present rates of supply of dissolved carbonates from the continents would provide 0·34 g/cm²/10³ yr of calcareous deposits over the entire ocean floor: two-thirds of this enters the Atlantic, however, and as a result the average $CaCO_3$ content of Atlantic sediments is 41%, compared with 19% in the Pacific. In the equatorial Pacific high-carbonate belt, accretion is at the rate of 0·01 m/10³ yr, compared with 0·0004–0·0005 m/10³ yr in the south-east Pacific. To understand what these rates mean, Menard has calculated that to cover the floor of the Pacific with a layer of beer cans 1 mm (0·001 m) thick over 10³ yr would require a supply of 10⁹ beer cans per day, equivalent to the world's entire output of steel.

Siliceous oozes, consisting of radiolarian tests and diatom frustules, cover 38×10^6 km². They accumulate in areas where terrigenous sediment supply is low, and where carbonates have been dissolved. Accumulation rates are about one-third of those for calcareous deposits.

Brown pelagic clays of mean particle size 1μ accumulate far from land, in the absence of both calcareous and siliceous material. Dating of the clay minerals has shown that the sediment source is terrigenous, including wind-transported desert dust and volcanic ash, together with meteoritic material. There is therefore some correlation with regional continental weathering patterns, and also a probability that accumulation rates were higher in the glacial periods. Brown clays, cover 102×10^6 km², mostly deeper than 4,500 m, and they are accumulating at the rate of $0 \cdot 0005 - 0 \cdot 0006$ m/10^3 yr; rates are higher in the north Pacific than in the south, and on the abyssal floor near Baja California reach $0 \cdot 009 - 0 \cdot 02$ m/10^3 yr.

The main diagenetic deposit on the deep sea floor consists of manganese nodules, composed of 16% Mn and 17·5% Fe. Mero finds the mean of 100 Pacific floor measurements to be 11 kg/m² of nodules, with a total volume on the floor estimated to be $1 \cdot 6 \times 10^{12}$ metric tons: manganese nodules are thus one of the most common rocks of the surface of the lithosphere. It is generally believed that nodule accumulation is taking place extremely slowly by accumulation of authigenic manganese at a rate of $0 \cdot 0003$ m/10^6 yr, but in this case nodules should be soon buried by pelagic deposits; recently it has been suggested that nodules form very rapidly following sporadic underwater volcanic effusion. At the slow rate nodules are being produced at the rate of 6×10^6 metric tons/yr.

Assuming a mean rate of pelagic deposition of $0 \cdot 003$ m/10^3 yr over a total area of 268×10^6 km², Poldervaart has calculated a total sediment accretion of $0 \cdot 8$ km³ or 22×10^8 tons/yr. This is equivalent to 6% of the present sediment load of rivers.

(b) Terrigenous deposition

Terrigenous deposits consist of muds (silty clays), which are green, black, or red, depending on chemistry and depositional conditions; glacial marine ice-rafted deposits in higher latitudes; and turbidites and slide deposits. Turbidite deposits of graded sands have been found over wide areas, related to local topography. They are also important in deep ocean trenches. Trenches more than 6,000 m deep (maximum depth 11,500 m) are concentrated in the Pacific, where they parallel continental shores and thus serve as sediment traps for terrigenous material. In the Puerto Rico trench the top 10 m of sediments consists of fine sands and silts, interpreted as turbidity current deposits: one such flow 3 m thick covers 10^4 km² and contains 3×10^7 m³ of sediment. In this case it is suggested that the entire $1 - 1 \cdot 7$ km of sediments on the trench floor may be Pleistocene turbidites. Other trenches contain little sediment, perhaps because they have formed so recently.

Other terrigenous deposits are important in smaller seas and basins such as the

Mediterranean and the Black Sea. Under euxinic conditions on the Black Sea floor seasonally varved clays and calcareous muds have accumulated at the rate of 1 m clay and 0·1–0·2 m calcareous mud in 5×10^3 yr. Rates in such nearshore basins tend to be highly variable and difficult to generalize.

(c) Biohermal deposits

Since the sea is approximately saturated with calcium carbonate, deposition of carbonates must be occurring in the ocean basins at the rate at which they are being supplied from the lands. This occurs mainly in the slow accumulation of pelagic carbonates over vast areas, and quantitatively the amount contained in bioherms (mainly in atolls, barrier, and fringing reefs) is minor. From borings on open-ocean atolls, we know that reef limestones have accumulated at the rate of 0·023 m/10^3 yr over the last 70×10^6 yr.

3. Erosion and sedimentation in the geologic record

The evidence of unconformities in the geologic record was used by James Hutton and both Powell and Dutton to infer prolonged periods of erosion, and such data have been used more recently to calculate the time required for peneplanation. This assumes that erosion rates have been invariant through time, a conclusion which stratigraphical geologists have long tended to dispute. Table 1.III.8 gives mean rates of sedimentation since the early Paleozoic calculated by Joseph Barrell; these show an apparent doubling in this time.

TABLE 1.III.8 Sedimentation rates in geologic time

Period	Maximum known thickness of strata (m × 10^3)	Approximate duration (yrs × 10^6)	Rate (m/10^3 yr)
Cenozoic	23·2	70	0·375
Mesozoic	33·2	120	0·299
Late Palaeozoic	27·4	130	0·187
Early Palaeozoic	29·0	180	0·161

Source: Joseph Barrell and A. Holmes, from Pettijohn [1957]

Menard has studied rates of sedimentation based on the geometry of deposits in three geosynclinal areas (the Appalachians, the Mississippi basin, and the Himalayas), in all of which the limits of the closed denudation system are inadequately known. His data, and calculated rates of denudation (which ignore dissolved load), are given in Table 1.III.9. Menard concludes that over 10^7–10^8 yr past erosion rates have varied by only one order of magnitude, compared with much greater variability at the present day. Whether such rates are representative of past conditions must depend on the extent and synchroneity of geosynclinal conditions, and hence on tectonic history. Gilluly has made a more detailed study of the Atlantic coast sediments of North America. He finds the Triassic and younger rocks to have an area of 490×10^3 km^2 and volume of $1,170 \times 10^3$ km^3, representing a volume of bedrock eroded of $1,023 \times 10^3$ km^3. The source area is delimited rather arbitrarily with an area of $1,320 \times 10^3$ km^2. Over the 225×10^6 yr of post-Triassic time the mean erosion rate has thus been

TABLE 1.III.9 Past erosion rates from depositional records

Area	Area eroded (km$^2 \times 10^6$)	Volume of deposits (km$^3 \times 10^6$)	Age (yr $\times 10^6$)	Total denudation (km)	Rate (m/10^3 yr)
Appalachian	1	7·8	125	7·8	0·062
Mississippi	3·2	11·1	150	6·9	0·046
Himalayas	1	8·5	40	8·5	0·21

Source: Menard [1961]

0·0034 m/10^3 yr. Taking the present suspended sediment load and dissolved load of the St Lawrence and other North Atlantic rivers, and making allowance for bed load, Gilluly calculated the present erosion rate to be 0·0218 m/10^3 yr, an order of magnitude greater. Present rates projected into the past would have supplied 60% more sediment than actually is found. These figures must be highly approximate, because of difficulties in delimiting the system; they also ignore reworking of sediments; but they do suggest that present rates are more rapid than those in the past, at least as recorded in basins of sedimentation.

Since long-continued sedimentation only occurs in subsiding depositional basins, inference from such data to past world erosion conditions must depend on the frequency, periodicity, and magnitude of basin development in the past. Thus the modern Mississippi delta is only the top member of a pile of sediments in the Gulf Coast geosyncline, accumulating since the Jurassic. During the Tertiary, contemporary with orogeny in western North America, 7,600 m of sediments formed in Louisiana and Texas, with a volume in the Paleogene of $1,000 \times 10^3$ km^3 and in the Neogene of 420×10^3 km^3. One deep well in south Louisiana, penetrating 6,880 m, reached only the Miocene. Subsidence is so rapid, partly from compaction, partly from deeper causes, that when rapid sedimentation ceases at the present delta, as the locus of deposition swings from side to side, channels are carried beneath the sea: dead deltas are revealed by the parallel islands of drowning levees. It was formerly thought that depression was caused by sediment-loading on the crust, and this certainly occurs, but to a limited extent. Thus at Lake Mead, where 40×10^9 tons of water and (so far) 2×10^9 tons of sediment have been added to 600 km^2 of crust, subsidence is taking place at the rate of 12·2 m/10^3 yr, but will total only 0·25 m.

World-wide geologic rates will also depend on the periodicity of orogeny: whether episodic, as argued by Stille and Umbgrove, or random in space and time, and hence essentially continuous, as argued by Shepard and Gilluly. In this connection present erosion rates may be inflated because of current tectonic activity: surprisingly high rates of earth movement have been revealed by geodetic survey, not only in mobile belts (up to 75 m/10^3 yr) and isostatic up-warps (5–15 m/10^3 yr) but in so-called stable areas like the Russian plains (up to 10 m/10^3 yr). Since we do not yet know tectonic norms for the world, we cannot assume that present conditions represent an erosion norm, especially when we take into account the major factors of Pleistocene glaciation, and the great amounts of unconsolidated sediments it left for later stream removal, and man's influence, in clearing the woodlands and cultivating the soil. These two factors alone could have inflated present river erosion rates by perhaps two orders of magnitude by comparison with a hypothetical norm.

Perhaps the most remarkable fact about sedimentation in the oceans is, over vast areas, its extreme slowness, and the concentration of most sediment into a few depositional basins. However, if present rates of 0·001–0·01 m/10^3 yr are extrapolated into the past, assuming the ocean basins to be 2 × 10^9 yr old, sediment thicknesses should be 1–10 km, nearer the latter than the former. Seismic work in the deep oceans has shown that the mean depth of unconsolidated material on the floors is about 0·5 km in the Pacific, 0·1–3 km in the Arctic, 0·6 km in the Indian Ocean, and 0·3–0·6 km in the Atlantic (down to zero on the mid-ocean ridge). No deep ocean sediments known are older than the Cretaceous. This suggests that ocean basins may be younger than was thought. Alternatively, the second layer in the crust (1·7 km thick) may be consolidated sediment rather than basalt. This problem of the absence in the ocean basins of sediments supplied from the continents has been called 'the great paradox of marine sedimentation' by Menard. Several workers, including Dietz, have speculated that ocean-floor sediments are re-incorporated in the continents, pelagic sediments floating on the tops of convection cells across the ocean floors from the mid-ocean ridges to the continents, where they sink beneath the sial layer.

Such speculations serve to illustrate the paucity of our knowledge on the long-term equilibrium of the world erosion–sedimentation system. At least it can be said that, with James Hutton, we see 'no vestige of a beginning, no prospect of an end'.

REFERENCES

ARRHENIUS, G. [1963], Pelagic sediments; In M. N. Hill, Editor: *The Sea*, volume 3 (New York), pp. 655–727.

CONALLY, J. R. and EWING, M. [1967], Sedimentation in the Puerto Rico Trench; *Journal of Sedimentary Petrology*, **37**, 44–59.

CORBEL, J. [1964], L'érosion terrestre, étude quantitative (Méthodes – techniques – résultats); *Annales de Géographie*, **73**, 385–412.

DOLE, R. B. and STABLER, H. [1905], Denudation; *United States Geological Survey, Water Supply Paper*, 294, 78–93.

DOUGLAS, I. [1967], Man, vegetation and the sediment yield of rivers; *Nature*, **215**, 925–8.

FOURNIER, F. [1960], *Climat et érosion: la relation entre l'érosion du sol par l'eau et les précipitations atmosphériques* (Paris), 201 p.

GIBBS, R. J. [1967], The geochemistry of the Amazon River system, Part I; *Bulletin of the Geological Society of America*, **78**, 1203–32.

GILLULY, J. [1949], Distribution of mountain building in geologic time; *Bulletin of the Geological Society of America*, **60**, 561–90.

GILLULY, J. [1964], Atlantic sediments, erosion rates, and the evolution of the continental shelf: some speculations; *Bulletin of the Geological Society of America*, **75**, 483–92.

HAYES, M. O. [1967], Relationship between coastal climate and bottom sediment type on the inner continental shelf; *Marine Geology*, **5**, 111–32.

HOLEMAN, J. N. [1968], The sediment yield of major rivers of the world; *Water Resources Research*, **4**, 737–47.

JUDSON, S. and RITTER, D. F. [1964], Rates of regional denudation in the United States; *Journal of Geophysical Research*, **69**, 3395–401.

KUENEN, P. H. [1946], Rate and mass of deep-sea sedimentation; *American Journal of Science*, **244**, 563–72.

KUENEN, P. H. [1950], *Marine Geology* (New York), 568 p.

MENARD, H. W. [1961], Some rates of regional erosion; *Journal of Geology*, **69**, 154–61.

MENARD, H. W. [1964], *Marine Geology of the Pacific* (New York), 271 p.

MENARD, H. W. and SMITH, S. M. [1966], Hypsometry of ocean basin provinces; *Journal of Geophysical Research*, **71**, 4305–25.

PETTIJOHN, F. J. [1957], *Sedimentary Rocks* (New York), 718 p.

POLDERVAART, A., Editor [1955], Crust of the earth; *Geological Society of America Special Paper*, **62**, 1–762.

SCHUMM, S. A. [1963], The disparity between present rates of denudation and orogeny; United States Geological Survey; *Professional Paper*, 454–H, 1–13.

SCRUTON, P. C. [1960], Delta building and the deltaic sequence; In Shepard, F. P., Phleger, F. B., and Van Andel, Tj. H., Editors, *Recent Sediments, Northwest Gulf of Mexico* (Tulsa), pp. 82–102.

SHEPARD, F. P. [1963], *Submarine Geology*; 2nd edn. (New York), 557 p.

SHEPARD, F. P., PHLEGER, F. B., and VAN ANDEL, TJ. H., Editors, *Recent Sediments, Northwest Gulf of Mexico* (Tulsa), 394 p.

STRAKHOV, N. M. [1967], *Principles of Lithogenesis*; volume 1 (London), 245 p.

2.II. The Drainage Basin as the Fundamental Geomorphic Unit

R. J. CHORLEY

Department of Geography, Cambridge University

1. Morphometric units

The need for the precise description of the geometry of landforms, particularly those of dominantly fluvial erosive origin, has been a recurring theme in geomorphology, and one of the most important aspects of this has been the search for the basic areal unit within which these data could be collected, organized, and analysed. The conceptions of the nature of these units have been very much a product of the broader methodological approaches to geography and earth science in general, and to geomorphology in particular, and can be grouped into three categories. The first important approach (Fenneman, 1914) sprang from the interest of geographers a half century ago in regional delimitation. The physiographic regions so delimited for the United States were based largely upon considerations of structural geology (e.g. the Ridge and Valley province), although certain gross morphometric attributes, notably relief and degree of dissection, were also used. The modern equivalent of this approach is provided by the terrain analogues of the U.S. Corps of Engineers, who used four terrain factors (characteristic slope, characteristic relief, occurrence of steep slopes greater than $26 \cdot 5°$, and the characteristic plan profile involving the 'peakedness', areal extent, elongation, and orientation of topographic highs) to divide up the *gross landscape* of a region into *component landscapes* in a simple taxonomic manner (fig. 2.II.1). In contrast with this basis, the second approach was concerned to identify 'the physiographic atoms out of which the matter of regions is built' (Wooldridge, 1932, p. 33). These 'atoms', however, were defined as the *facets* of 'flats' and 'slopes' forming the intersecting surfaces characteristic of polycyclic landscapes (Wooldridge, 1932, pp. 31–3), and, although this doctrinaire definition has been relaxed to include *segments* of smoothly curved surface (Savigear, 1965) and to allow the grouping of facets into *landscape patterns*, such as a 'mature river valley' (Beckett and Webster, 1962) (fig. 2.II.1), the genetic overtones and subjective character of this morphometric division limits its usefulness (Gregory and Brown, 1966). The third basis for morphometric division results from the obvious unitary features both of geometry and process exhibited by the erosional drainage basin, as recognized long ago by Playfair (Chorley, Dunn and Beckinsale, 1964, pp. 61–3) and by Davis [1899], who wrote:

Although the river and the hill-side waste sheet do not resemble each other at first sight, they are only the extreme members of a continuous series, and when this generalization is appreciated, one may fairly extend the 'river' all over its basin and up to its very divides. Ordinarily treated, the river is like the veins of a leaf; broadly viewed, it is like the entire leaf.

This topographic, hydraulic, and hydrological unity of the basin provided the basis for the morphometric system of R. E. Horton [1945], as elaborated by Strahler [1964], and it is now employed as a basic erosional landscape element because it is:

1. A limited, convenient, and usually clearly defined and unambiguous topographic unit, available in a nested hierarchy of sizes on the basis of stream
ordering.
2. An open physical system in terms of inputs of precipitation and solar
radiation, and outputs of discharge, evaporation, and reradiation (Lee,
1964).

2. Linear aspects of the basin

The defining of the perimeter of a drainage basin in the above terms is not difficult, especially as in the majority of instances the ground-water divides are coincident with the topographic ones, but more problems are presented in the definition of the stream-channel network. Definition is especially difficult for the fingertip tributaries in regions of deep soil and plentiful vegetation, whereas in arid shale badlands it is also difficult to distinguish between permanent channels infrequently occupied by runoff and ephemeral rills. The definition of a stream segment, either from the map or in the field, involves five considerations: for a given region or map scale it must have a lower limiting size; it must be con-nected with the main stream network; it must be 'permanent', as distinct from seasonal; it must form part of a distinctly bifurcating channel pattern; and it must conduct laterally concentrated surface runoff from a well-defined drainage area. A further problem is that the heads of distinct channels are constantly migrating in response to storm excavation or prolonged infill of slope debris (Kirkby and Chorley, 1967).

The linear aspects of stream networks can be analysed from two main view-points:

(a) the *topological*, which considers the interconnections of the system and
yields some scheme of stream ordering; and
(b) the *geometrical*, having to do with the lengths, shapes, and orientations of
the constituent parts of the network.

The recognition of a hierarchy of stream segments is important because of the different morphometric and hydrologic features associated with each. The most widely used ordering scheme was adapted by Strahler [see, for example, 1964] from Horton [1945], in which fingertip channels are specified as order (U) 1, and where two first-order tributaries join, a channel segment of order 2 is formed,

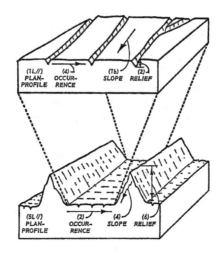

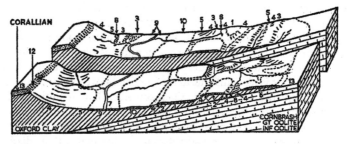

Fig. 2.11.1 Landscape units and geomorphic regions.

Above: Example of a component landscape defined in terms of four terrain factors, and the relation between a component (*top*) and a gross landscape (From Van Lopik and Kolb, 1959).

Below: The pattern of a mature river valley developed by the Upper Thames on the Oxford Clay, illustrating the facets and their relation to each other in the landscape (From Beckett and Webster, 1962).

Facets of river valley and clay

1. High gravel terrace.
3. Clay crest.
5. Clay footslope.
7. River and banks.
9. Flood plain alluvium.

2. Spring line.
4. Clay slope.
6. Unbedded glacial drift.
8. Local bottomland.
10. Old alluvium, not flooded.

Facets of scarplands bounding the river valley pattern

12 (11) Scarp slope.

13. Dipslope.

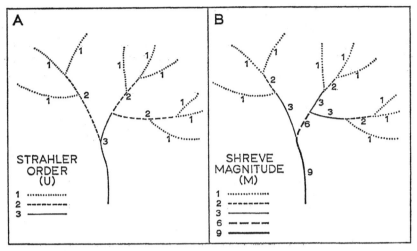

Fig. 2.11.2 Two methods of stream network ordering: (A) Stream segment orders (After Strahler); (B) Stream link magnitudes (After Shreve).

etc. (fig. 2.11.2(*a*)). The main disadvantage of this Strahler system is that it violates the distributive law, in that the entry of a lower-order tributary stream does not always increase the order of the main stream, and Shreve [1966] has proposed a simple remedy for this by dividing the network into separate links at each junction and allowing the magnitude of each link to reflect the number of first-order fingertips ultimately feeding it (fig. 2.11.2(*b*)), and other more involved

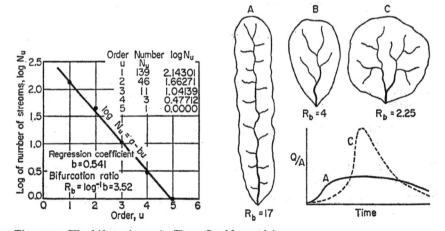

Fig. 2.11.3 The bifurcation ratio (From Strahler, 1964).
Left: Plot of number of stream segments versus order, with a fitted regression.
Right: Hypothetical drainage basins of differing bifurcation ratios, together with their extreme effects on the runoff hydrograph.

schemes have been suggested. However, the simpler unambiguous Strahler system is now firmly established, and this ordering system provides sequences of stream order numbers $(N_1, N_2, \ldots N_K)$ which approximate an inverse geometric series for a given basin with the degree of branching, or bifurcation ratio (R_b), given by the ratios N_1/N_2, N_2/N_3, etc., or the antilog of

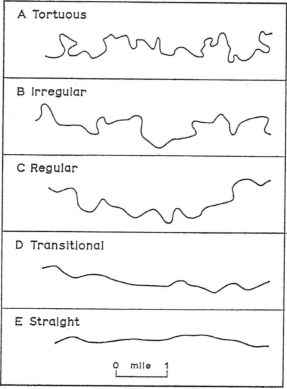

A Tortuous

B Irregular

C Regular

D Transitional

E Straight

0 mile 1

Fig. 2.11.4 Examples of channel patterns (From Schumm, S.A., 1963, *Bulletin of the Geological Society of America*). (A) White River near Whitney, Neb. ($P = 2\cdot1$); (B) Solomon River near Niles, Kan. ($P = 1\cdot7$); (C) South Loup River near St. Michael, Neb. ($P = 1\cdot5$); (D) North Fork Republican River near Benkleman, Neb. ($P = 1\cdot2$); (E) Niobrara River near Hay Springs, Neb. ($P = 1\cdot0$).

the regression coefficient (b) (fig. 2.11.3). The bifurcation ratio, for a given density of drainage lines, is very much controlled by basin shape and shows very little variation (ranging between 3 and 5) in homogeneous bedrock from one area to another. Where structural effects cause basin elongation, however, this value may increase appreciably. Besides influencing the landscape morphometry, the bifurcation ratio is an important control over the 'peakedness' of the runoff hydrograph (fig. 2.11.3) (see Chapter 9.1).

The ratio between the measured length of a stream channel and that of the

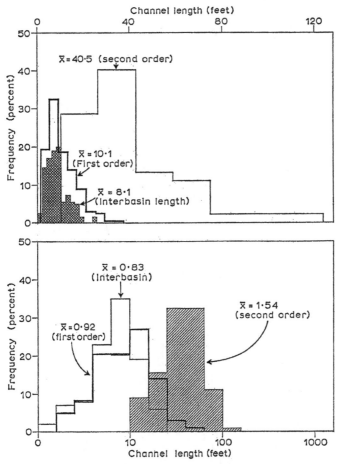

Fig. 2.11.5 Frequency-distribution histograms of first- and second-order channel lengths and maximum interbasin lengths for the Perth Amboy Badlands, New Jersey (From Schumm, 1956).

Above: Actual stream lengths.
Below: Logarithms of stream lengths.

thalweg of its valley is a measure of its sinuosity (fig. 2.11.4). Distributions of lengths of streams of each order in a drainage basin are characteristically right-skewed (log-normal) (fig. 2.11.5), and the plot of mean stream lengths of each order ($\bar{L}_1$, $\bar{L}_2$, $\bar{L}_3$, ... $\bar{L}_K$) in a basin produces an approximation to a direct geometric series (fig. 2.11.6), where the antilog of the regression coefficient is the length ration (R_l).

Obviously the absolute length of the channel system exercises a strong control over the basin lag time (the time difference between rainfall and the resulting

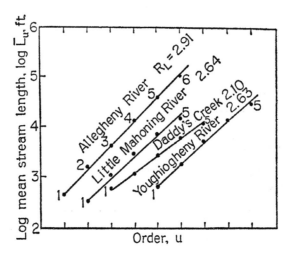

Fig. 2.11.6 Regression of logarithm of mean stream segment length versus order for four drainage basins in the Appalachian Plateau Province (After Morisawa; From Strahler, 1964).

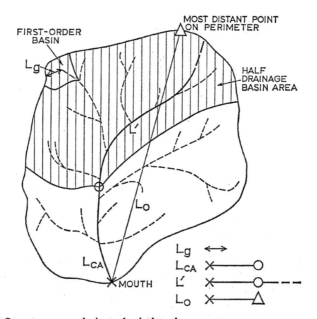

Fig. 2.11.7 Some common drainage basin length parameters.

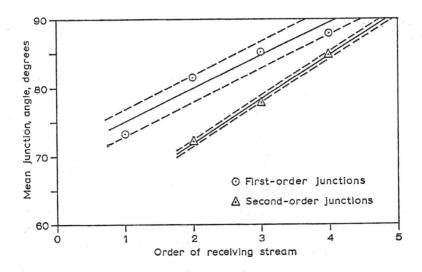

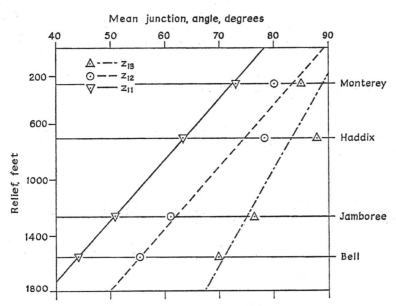

Fig. 2.11.8 Relationships of mean stream junction angles (From Lubowe, J. K., 1964, *American Journal of Science*).

Above: Plot of mean junction angle of first- and second-order with receiving streams of similar or higher order, in part of the Lexington Plain, Kentucky.

Below: Plot of mean junction angles of first-order streams with first-, second-, and third-order receiving streams in four areas of the east, central and western United States.

stream runoff: see Chapter 9.1), as do the following length parameters (fig. 2.II.7):

1. The length of overland flow (L_g), which is the mean horizontal length of the flow path from the divide to the stream in a first-order basin. This parameter is a measure of stream spacing, or degree of dissection, and is approximately one-half the reciprocal of the drainage density $\left(L_g \simeq \dfrac{1}{2D}\right)$. As the mean velocity of unconcentrated overland flow is less than $\frac{1}{5}$ that of concentrated channel flow (30,000 cm/hr, versus 160,000 cm/hr), L_g is also an important control over lag time. A less-meaningful measure of stream spacing is the mean stream interval (MI), calculated from sampling stream intersections with a grid.

2. The length of the main stream of the basin (L') (usually designated by following up from the mouth those streams which make the least angle with the next lower segment). This is sometimes continued to the basin margin (then called the 'mesh length').

3. The distance to the 'centre of gravity' of the drainage basin (L_{ca}). This is usually measured up the main stream to a point where one half of the drainage basin area lies headward of it, but for most basins $L_{ca} = 0.5L$ is a good approximation. It is sometimes measured to the centre of gravity of the basin area.

4. The length of the longest basin diameter (L_0), measured from the basin mouth to the most distant point on the perimeter. This is useful in the calculation of basin shape.

The entrance angle (Z_c) between a tributary developed in a valley-side slope (of $\theta°$) and joining a larger stream of lower slope ($\gamma°$) would be approximately expressed by:

$$\cos Z_c = \frac{\tan \gamma}{\tan \theta}$$

As the slopes are degraded through time, one might expect θ to approach γ, and consequently Z_c to decrease. It is characteristic that mean values of junction angle increase as the order of the receiving stream increases (i.e as the difference between γ and θ increases), and it is inversely related to relief (for given orders of junction), probably because high relief imparts especially high gradients to the receiving streams (fig. 2.II.8).

3. Areal aspects of the basin

Basin area (conventionally referred to a horizontal datum plane) is hydrologically important because it directly affects the size of the storm hydrograph and the magnitudes of peak and mean runoff (fig. 2.II.9) (See Chapter 9.1). It is interesting that the maximum flood discharge per unit area is inversely related to size, because the most intense storms are usually of the smallest size (fig. 2.II.9) (More, 1967, p. 166). In a given large drainage basin developed in a homogeneous region

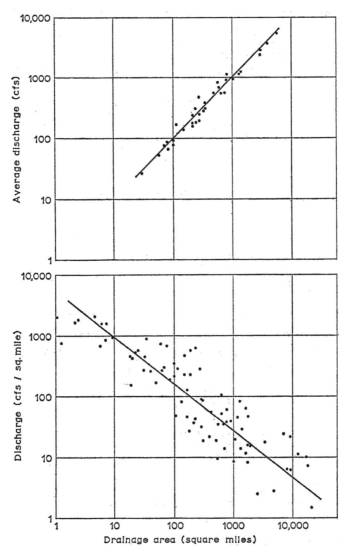

Fig. 2.11.9 Relations between basin area and stream discharge.

Above: Mean discharge (cfs) versus drainage area for all gauging stations on the Potomac River (After Hack: From Strahler, 1964).

Below: Maximum flood discharge (cfs) per square mile versus drainage area for basins in Colorado (From Follansbee, R. and Sawyer, L. R., 1948, *U.S. Geological Survey Water Supply Paper* 997).

basin areas of given order show a logarithmic-normal distribution (fig. 2.II.10), the means of which approximate a direct geometric series (fig. 2.II.11). By relating characteristic discharge to drainage area (the relationship $Q_{2\cdot33} = 12A^{0\cdot79}$ was obtained for basins in central New Mexico, where $Q_{2\cdot33}$ is the flood discharge equalled or exceeded on average once every 2·33 years, or the mean annual

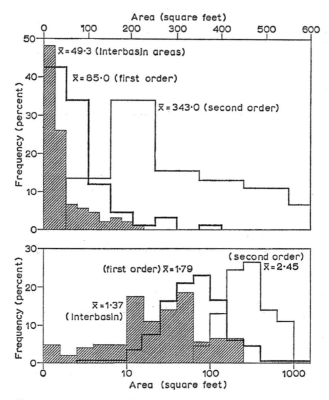

Fig. 2.II.10 Frequency-distribution histograms of first- and second-order basin areas and interbasin areas for the Perth Amboy Badlands, New Jersey (From Schumm, 1956).

Above: Actual basin areas.
Below: Logarithms of basin areas.

flood: see Chapter 7.II), and then area to order, it is possible to show an exponential relationship between order and discharge (fig. 2.II.12).

The relationship of stream length to basin area is important because plots of basin area draining into various locations along the main stream (i.e. area-distance curves) give an idea of the pattern of runoff (fig. 2.II.13), and also because the relationship of total stream lengths of all orders to basin area is one of the most sensitive and variable morphometric parameters, and one which controls the texture of landscape dissection and the spacing of streams. Thus drainage density (D), for example, is defined as the total stream length per unit area of basin (e.g. in miles per square mile). Drainage density exhibits a very wide range of values in nature and is commonly believed to reflect the operation of the complex factors controlling surface runoff. Common values of D are, for example, 3–4 in the sandstones of Exmoor and the Appalachian plateaus,

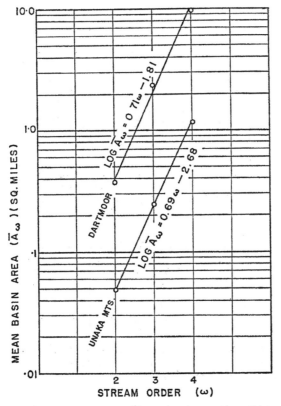

Fig. 2.11.11 Regressions of mean area versus order for drainage basins in the Unaka Mountains, Tenn. and N. Car., and Dartmoor, England (From Chorley, R. J. and Morgan, M. A., 1962, *Bulletin of the Geological Society of America*).

20–30 for the scrub-covered Coast Ranges of California, 200–400 for the shales of the Dakota badlands, and up to 1,300 for unvegetated clays (fig. 2.11.14). An allied measure is the constant of channel maintenance which is the area (in square feet) necessary to maintain 1 ft of drainage channel. Drainage density affects the runoff pattern, in that a high drainage density removes surface runoff rapidly, decreasing the lag time and increasing the peak of the hydrograph (see Chapter 9.1). Of course, basin shape (itself largely controlled by geological structure) is an important control over the geometry of the stream network (fig. 2.11.15). Examples of simple shape measures are:

1. The circularity ratio – (see Chapter 9.1).
2. The elongation ratio – the diameter of a circle having the same area as the basin, as a ratio of the maximum basin length (L_0).
3. The measure $(L' . L_{ca})^{0.3}$, found by experience to be a good predictor of basin lag.

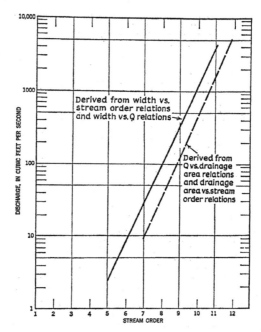

Fig. 2.11.12 Relation of discharge to stream order for ephemeral streams in New Mexico derived by two separate types of analysis (From Leopold, L. B. and Miller, J. P., 1956, *U.S. Geological Survey Professional Paper* 282–A).

It is interesting that, unless pronounced structural control is present, drainage basins differ relatively little in shape, although basins tend to become more elongate with strong relief and steep slopes.

4. Relief aspects of the basin

Longitudinal profiles of stream channels are characteristically concave-up, and it has been suggested that this is due to the increase of discharge downstream not being balanced by any commensurate increase in frictional losses – the increase in the wetted perimeter being more than compensated for by the increase in cross-sectional area and by the decrease of bed material grain size (due to sorting and abrasion during transportation). Many attempts have been made to fit

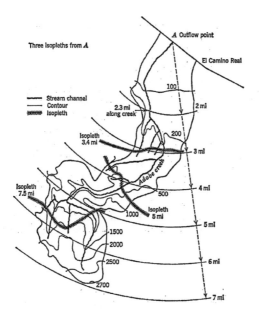

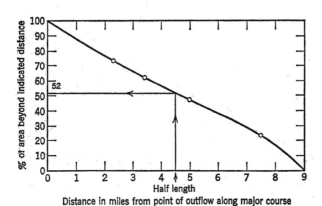

Fig. 2.II.13 Area as a function of channel distance from the basin mouth for Adobe Creek, near Palo Alto, California (Area 11 square miles; drainage density 2·18 miles/square mile) (From De Wiest, 1965).

Above: Adobe Creek, showing channel distance isopleths.
Below: Distribution of area as a function of distance from the basin mouth.

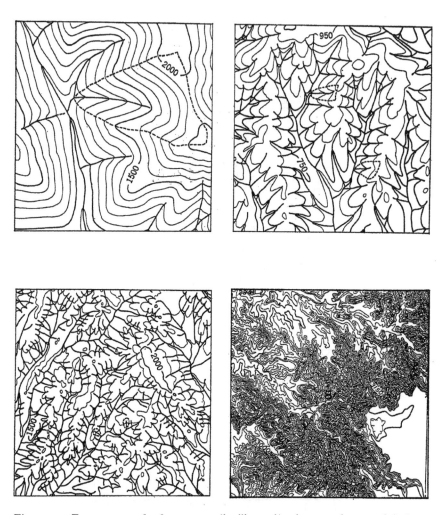

Fig. 2.11.14 Four areas, each of 1 square mile, illustrating the natural range of drainage density (From Strahler, 1964).

Top left: Low drainage density: Driftwood Quad., Penn.
Top right: Medium drainage density: Nashville Quad., Ind.
Bottom left: High drainage density: Little Tujungo Quad., Cal.
Bottom right: Very high drainage density: Cuny Table West Quad., S. Dak.

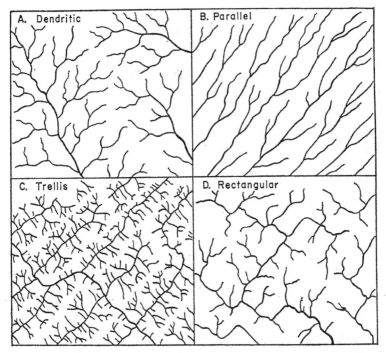

Fig. 2.11.15 Four basic drainage patterns, each occurring at a wide range of scales (From Howard, A. D., 1967, *Bulletin of the American Association of Petroleum Geologists*).

mathematical curves, chiefly logarithmic and exponential, to complete stream profiles, largely to extrapolate them in order to identify supposed ancient base levels. Even the exponential curve, which is most theoretically attractive because it also applies to the rate of attrition of transported debris, usually provides an imperfect fit, and long stream sections seem to be divided into segments by discontinuities which are due to changing discharge, calibre of bedload, or both, or to changing channel characteristics. The influence of discharge is shown by the segmentation on the basis of stream orders in fig. 2.11.16. It has been suggested that in a given basin mean channel gradient bears an inverse geometric relationship to stream order (fig. 2.11.17), although the imperfection of this relationship has led to other more complex ones being suggested. For practical purposes other stream-gradient measures are employed, including:

1. Some measure of the average slope of the main channel in a basin. This is commonly the arithmetic mean slope ($\bar{S}$) of the whole channel, or the slope of the 'equivalent' stream (i.e. one having the same length and flood peak travel time) (S_{st}).

2. The simple gross slope of the channel, obtained by dividing the elevation difference between head and mouth by the length of the main stream.

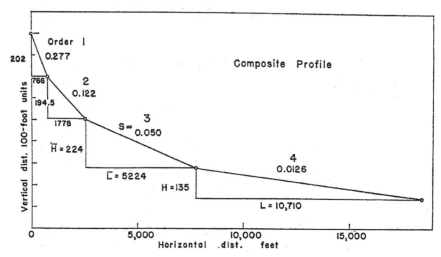

Fig. 2.11.16 Plot of the longitudinal profile of Salt Run, Penn., showing the difference in mean slope of each of the four segments of differing order (From Broscoe, 1959, *Office of Naval Research Technical Report 18, Project NR 389–042, Contract N6 ONR 271–30*).

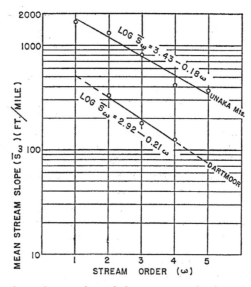

Fig. 2.11.17 Regressions of mean channel slope versus order for streams in the Unaka Mountains, Tenn. and N. Car., and Dartmoor, England (From Chorley, R. J. and Morgan, M. A., 1962, *Bulletin of the Geological Society of America*).

3. The mean slope of the whole channel system, computed by averaging the gradients of all channels draining at least 10% of the total basin area.

These slopes are important because channel slope exercises an important influence over the magnitude of the runoff peak (see Chapter 9.1).

Orthogonal ground-slope angles (S_g) in a basin are commonly measured at the maximum gradient of given valley-side profiles, or expressed as the mean angle of a complete valley-side profile, or sampled over the whole basin. Both the

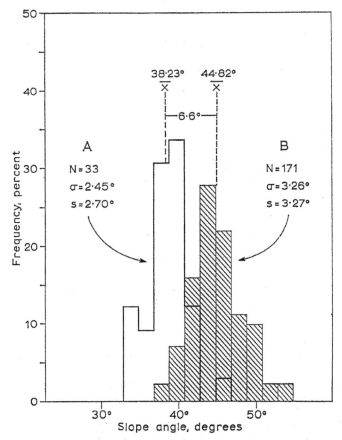

Fig. 2.11.18 Comparison of maximum valley-side slope angles in the Verdugo Hills, California, (A) protected at the base by talus and slope wash, and (B) actively corraded at the base (From Strahler, A. N., 1954, *Journal of Geology*).

maximum and mean valley-side slope angles within a basin are commonly normally distributed (fig. 2.11.18), and, although they are not individually related to the gradient of the basal stream, a mean relationship seems to exist when different regions are compared. The character of the distribution of slope angles sampled over the whole basin depends on the height distribution within it; for a 'just mature' basin with limited flat summit or floodplain areas the distribution is normal, but other basins give skewed distributions, the direction of skew depending on whether the small angles are concentrated on the summits ('youth') or on the floodplains ('late maturity'). Again, simple measures of average ground

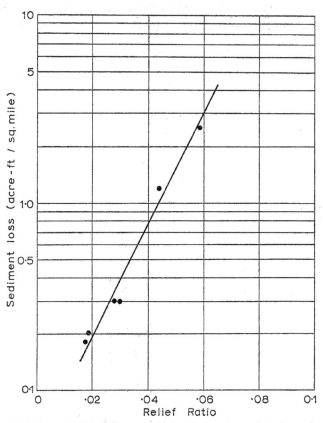

Fig. 2.II.19 Relation of sediment loss to relief ratio for six small drainage basins in the Colorado Plateaus (From Schumm, 1956).

slope within a basin are employed in hydrological analysis, such as the mean basin slope $\left(\dfrac{\text{Total length of contours} \times \text{Contour interval}}{\text{Basin area}}\right)$, and the relief ratio (R_h), which is the ratio of the maximum basin relief to the horizontal distance along the longest basin dimension parallel to the main drainage line. Even such crude measures as these can be used to rationalize basin dynamics, it being found that mean basin slope influences the form of the hydrograph and that the relief ratio exercises an important control over rates of sediment loss from some basins (fig. 2.II.19).

The distributional characteristics of land elevations have long been of geomorphic interest, because concentrations of area with elevation (i.e. surfaces of low slope) were believed to be indicative of ancient base levels. Plots of mean land slope versus elevation (clinographic curves) and of amount of surface area versus elevation (hypsometric curves) were prepared for large upland areas to assist in the evaluation of their possible polycyclic histories. Both techniques have been re-applied to individual drainage basins; a clinographic curve giving a

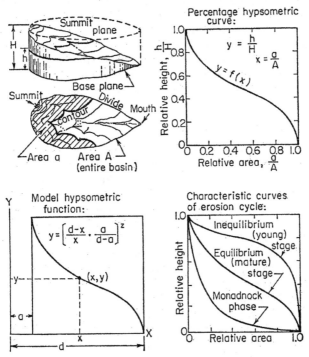

Fig. 2.11.20 The calculation of the hypsometric curve (From Strahler, A.N., 1957, *Transactions of the American Geophysical Union*).

more accurate estimate of land slope than average figures when evaluating the form of the hydrograph, and the dimensionless hypsometric curve representing in some instances the relative stage of basin degradation through time with reference to an assumed original uneroded block (fig. 2.11.20). An especially important hydrological property of the basin related to the distribution of elevations is the amount of floodplain storage available, the effect of which is to make the rising limb of the hydrograph less steep, increase the lag time, and make the peak lower and less pronounced. A knowledge of the distribution of elevations also enables better estimates of rainfall, snowfall, and evaporation in the basin to be made.

In the past, morphometric analysis from maps has been a rather tedious and time-consuming task, but recently techniques have been devised to give it greater facility. The most important of these is the digitizer of the 'pencil-follower' type, which both records the rectangular coordinates of points on a map and also gives a continuous read-out of point coordinates sufficiently closely spaced (maximum recording rate is 20 points per second) to define lines – i.e. contours, stream channels, basin perimeters, etc. The card or tape output can be edited, elaborated, and then fed directly into a computer, together with programmes which will automatically calculate areas, shapes, drainage densities, mean aximuths, maximum slope angles, and the like. Such programming is in its infancy, but already it promises to release the masses of data locked up in topographic maps and will obviously allow much more extensive sampling and generalization of morphometric properties. Before too long these methods will be applied directly to the output from aircraft and satellite scanning equipment, obviating the necessity for the actual compilation of many maps.

5. Some considerations of scale

Considerations of changing scale, both linear and temporal, introduce a certain element of sophistication into studies concerned with morphometric relations.

Similarities such as appear to exist generally between the bifurcation, length, and area ratios of basins of differing size developed on bedrock lacking pronounced structural control have prompted speculation that erosional drainage basins in differing hydrological environments may show a close approximation to geometrical similarity when mean values are considered. If complete geometrical similarity existed one would expect to find all length measurements between corresponding mean points to bear a fixed linear scale ratio, and all corresponding angles to be equal. Although this does not seem to be the case for all morphometric properties (i.e. there appears to be a changing relationship between main-stream length and basin area as basins increase in size within a region, which seems due to both an increasing elongation of larger basins and to the increasing sinuosity of the stream), the similar geometrical relationships between some linear properties suggest a more general significance to the 'laws of morphometry'.

Although it has not been the purpose of this contribution to discuss the relationships between morphometry and erosional processes, it has doubtless become apparent that two quite distinct approaches to this are possible, dependent upon the time-scale with which one is concerned. In the long term it is clear that the hydrological events which compose 'process' must be instrumental in determining the morphometry of the landscape, but, when one is concerned with explaining in the short term the factors which control the character of such individual processes, morphometric features (such as gradient) are commonly invoked. Whether morphometric parameters are viewed as mathematically dependent or independent variables is very much a matter of the time scale employed. Figure 2.11.21 gives two plots involving drainage density, one showing it to be strongly controlled by the precipitation effectiveness and the other

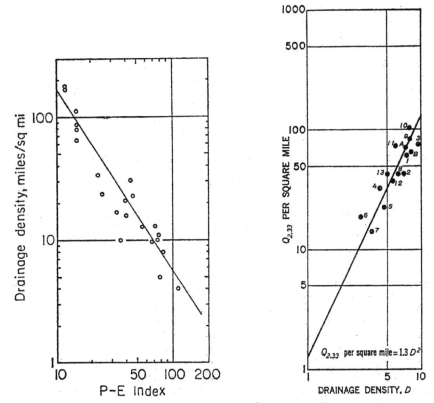

Fig. 2.11.21 The control of drainage density exercised by Thornthwaite's precipitation effectiveness (P–E) index (*Left*) (From Melton, M. A., 1957, *Office of Naval Research Technical Report 16, Project NR 389–042, Contract N6 ONR 271–30*); and (*Right*) the control exercised by drainage density over the mean annual flood ($Q_{2.33}$) for 13 basins in the central and eastern United States (From Carlston, C. A., 1963, *U.S. Geological Survey Professional Paper 422–C*).

indicating how drainage density differences within a region control the peak mean annual flood. A number of the following chapters examine the effects of hydrological processes on aspects of basin morphometry, and Chapter 9.1 uses basin morphometry, among other factors, to analyse the form of the flood hydrograph.

REFERENCES

BECKETT, P. H. T. and WEBSTER, R. [1962], The storage and collection of information on terrain (An interim report); *Military Engineering Experimental Establishment, Christchurch, Hampshire*, 39 pp. (Mimeo).

CHORLEY, R. J., DUNN, A. J., and BECKINSALE, R. P. [1964], *The History of the Study of Landforms*, Volume I (Methuen, London), 678 pp.

CLARKE, J. I. [1966], Morphometry from maps; In Dury, G. H., Editor, *Essays in Geomorphology* (Heinemann, London), pp. 235–74.

DAVIS, W. M. [1899], The geographical cycle; *Geographical Journal*, 14, 481–504.

DE WIEST, R. J. M. [1965], *Geohydrology* (Wiley, New York), 366 pp.

FENNEMAN, N. M. [1914], Physiographic boundaries within the United States; *Annals of the Association of American Geographers*, 4, 84–134.

GOLDING, B. L. and LOW, D. E. [1960], Physical characteristics of drainage basins; *Proceedings of the American Society of Civil Engineers, Journal of the Hydraulics Division*, 86, No. HY 3, 1–11.

GRAY, D. M. [1961], Interrelationships of watershed characteristics; *Journal of Geophysical Research*, 66, 1215–23.

GREGORY, K. J. and BROWN, E. H. [1966], Data processing and the study of land form; *Zeitschrift für Geomorphologie*, Band 10, 237–63.

HORTON, R. E. [1945], Erosional development of streams and their drainage basins: Hydrophysical approach to quantitative morphology; *Bulletin of the Geological Society of America*, 56, 275–370.

KIRKBY, M. J. and CHORLEY, R. J. [1967], Throughflow, overland flow and erosion; *Bulletin of the International Association of Scientific Hydrology*, Year 12(3), 5–21.

LANGBEIN, W. B. *et al.* [1947], Topographic characteristics of drainage basins; *U.S. Geological Survey Water Supply Paper* 968-C, 125–157.

LEE, R. [1964], Potential insolation as a topoclimatic characteristic of drainage basins; *Bulletin of the International Association of Scientific Hydrology*, Year 9, 27–41.

LEOPOLD, L. B., WOLMAN, M. G., and MILLER, J. P. [1964], *Fluvial Processes in Geomorphology* (Freeman, San Francisco), pp. 131–50.

MORE, R. J. [1967], Hydrological models and geography; In Chorley, R. J. and Haggett, P., Editors, *Models in Geography* (Methuen, London), pp. 145–85.

SAVIGEAR, R. A. G. [1965], A technique for morphological mapping; *Annals of the Association of American Geographers*, 55, 514–38.

SCHUMM, S. A. [1956], The evolution of drainage systems and slopes in badlands at Perth Amboy, New Jersey; *Bulletin of the Geological Society of America*, 67, 597–646.

SHREVE, R. L. [1966], Statistical law of stream numbers; *Journal of Geology*, 74, 17–37.

STRAHLER, A. N. [1964], Quantitative geomorphology of drainage basins and channel networks; In Chow, V. T., Editor, *Handbook of Applied Hydrology* (McGraw-Hill, New York), Section 4–11.

VAN LOPIK, J. R. and KOLB, C. R. [1959], A technique for preparing desert terrain analogs; *U.S. Army Engineer Waterways Experiment Station, Vicksburg, Mississippi, Technical Report* 3–506, 70 pp.

WISLER, C. O. and BRATER, E. F. [1959], *Hydrology*; 2nd edn. (Wiley, New York), 408 pp.

WOOLDRIDGE, S. W. [1932], The cycle of erosion and the representation of relief; *Scottish Geographical Magazine*, 48, 30–6.

3.II. The Role of Water in Rock Disintegration

R. J. CHORLEY

Department of Geography, Cambridge University

1. Some properties of water

Water molecules are formed by the bonding of two hydrogen atoms to one oxygen atom as a result of the former sharing their single negatively charged electrons with the oxygen, giving the latter the optimum eight in its outermost shell (fig. 3.II.1(a)). This bonding is termed *covalent* and produces a molecule in which the hydrogen atoms have a net positive charge and the oxygen a double negative one. The asymmetrical bonding of the hydrogen atoms makes the molecule dipolar, in that one end is charged negatively and the other positively (fig. 3.II.1(b)). This has a number of consequences:

1. Under some circumstances the molecule can separate into two oppositely charged ions (H^+ and OH^-) which make it more available for some chemical reactions. OH^- is the hydroxyl which is important in many weathering reactions, and H^+ is the cation whose effective concentration imparts to an aqueous solution its acidity. The range of the latter is so large that acidity (pH) is expressed as the negative logarithm of the free H^+ concentration in grams per litre (ph = 7 is thus 0·0000001 g of free H^+ ions per litre of water; this is the neutral value, higher pH values indicate alkalinity, lower values acidity). The solution of atmospheric carbon dioxide by rain-water produces free H^+ ions in the resulting carbonic acid:

$$CO_2 + HOH \rightleftharpoons H^+ + HCO_3^-$$

2. By orienting themselves in an electrical field water molecules can weaken it. This explains the very effective solvent action of water when it is in contact with other molecules, the atoms of which are simply held together because the component ions are of opposite electrical charge (i.e. ionic bonding). The best example of this is the solution of common salt by the negative and positive ends of the water molecules attaching themselves respectively to the Na^+ and Cl^- ions, neutralizing their charges so that the least mechanical agitation will float the sodium and chlorine ions apart.

3. The positioning of the two hydrogen atoms means that the water molecule is basically four-cornered, with the negative charges concentrated at two points on the surface of the oxygen atom. When water molecules come into association, therefore, they tend to join together by ionic bonding (hydrogen

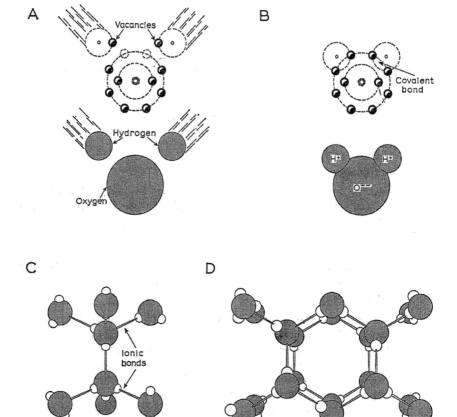

Fig. 3.II.1 The chemical structure of water (After Davis and Day, 1964).
A. The joining of two hydrogen with one oxygen atom.
B. The covalent bond in the water molecule.
C. The ionic bonding of water molecules.
D. The structure of ice.

bonding) into tetrahedral groups of four (fig. 3.II.1(c)). The positive charge of the hydrogen nucleus is attracted to the oxygen ion with a force which is only about 6% of that which binds the hydrogen to the oxygen atoms within the molecule. This phenomenon of hydrogen bonding gives rise to such water properties as surface tension and capillarity (i.e. cohesion), and to adhesion to some surfaces – particularly those having surface oxygen atoms (e.g. glass, quartz, clay minerals, etc.). It also explains the high theoretical tensile strength of water, which is about the same as some steels.

4. As a liquid water is atypical, in that it seems to be composed of clusters of molecules connected by hydrogen bonds, separated from one another by

unbound water molecules which can rotate freely, forming lubricating layers and allowing flowage (fig. 3.II.2). The structure of water is not static, and molecules exchange rapidly between the clusters and the flow layers and, on average, each intermolecular hydrogen bond breaks and reforms 10^{12} times a second. As the temperature of water decreases the thermal agitation of the molecules also decreases until the maximum density is reached at about $4°$ C, this increase of density, together with an increase in the number of hydrogen bonds (and therefore of cluster size), means that viscosity (or internal resistance to deformation) is also inversely related to

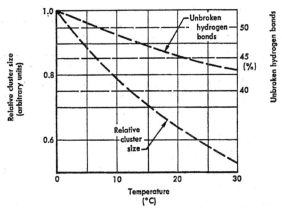

Fig. 3.II.2 The effect of temperature on the relative number of unbroken hydrogen bonds and the relative cluster size in pure water (Data from Nemethy and Scheraga. From Gross, M. G., *Oceanography*, Merrill Physical Science Series, 1967).

temperature (fig. 3.II.2). At $20°$ C the absolute viscosity of water is little more than half that at $0°$ C.

5. When the temperature of water falls to about $4°$ C the expansion due to the widespread formation of the open hydrogen bonds exceeds the contraction due to the decreasing molecular thermal agitation, and the water begins to expand in volume and assume a lower density as more and more of the tetrahedral clusters are taken up into a hexagonal structure (fig. 3.II.1(d)). This expansion continues until at a temperature of $-22°$ C ice achieves its minimum density and maximum expansive pressure under confining conditions. At lower temperatures ice some 3% more dense than water begins to form, and at $-70°$ C the crystal habit changes from hexagonal to cubic. In its usual hexagonal form cleavage is parallel to the basal plane, and the maximum growth rate is normal to this. The former property is of great importance in the flow properties of glacial ice, and the latter in its mechanical action in weathering. Except beneath a more or less flat water-table near enough to the surface to freeze, water in natural circumstances seldom freezes in completely confined conditions, but under conditions

which allow an outside water supply to the growing crystals, which tend to develop in clusters of parallel needles at right angles to the freezing surface (i.e. the air or a rock surface). Under these conditions the stresses developed by ice growth may be ten times those associated with the simple expansion of water during freezing, for they are limited only by the tensile strength of water, which is drawing the water molecules through the capillary films to the ends of the growing ice crystals.

2. Weathering of igneous minerals and rocks

The weathering of igneous rocks may be defined as the response of mineral assemblages which were crystallized in equilibrium at high pressure and temperature within the earth's crust to new conditions at or near contact with air, water, and living matter, giving rise to their irreversible change from the massive to the clastic or plastic state involving increases in bulk, decreases in density and particle size, and the production of new minerals more stable under the new conditions. In this process water plays a dominant role, which can only be understood by first examining the structure of the silicate minerals which compose virtually all of igneous rocks and make up more than 90% of all rock-forming minerals.

The basic building block of the silicate minerals is the silica tetrahedron (SiO_4), in which a silicon ion (Si^{++++}) fits snugly between four oxygen ions (O^{--}) covalently bonding them by sharing one electron with each (fig. 3.III.3(a)). This structure is both chemically efficient and geometrically compact, making the tetrahedra strong and very difficult to break up chemically. However, the silicate minerals are also composed of other positive metal ions (cations), such as aluminium (Al^{+++}), iron (Fe^{+++} or Fe^{++}), magnesium (Mg^{++}), calcium (Ca^{++}), sodium (Na^{+}), and potassium (K^{+}), and these form generally the weaker links in the crystalline structures. In particular, the smaller electrical charges of the latter (especially of Na^{+} and K^{+}) make them very susceptible to being neutralized by the dipolar activity of water if it can enter the crystal lattice. It is clear, then, that the most resistant minerals are those formed exclusively of interlocking silica tetrahedra, and the less the interlocking and the greater the interpolation of the other metal cations, the more readily the mineral will break down under the action of water. Apart from orthoclase (potash feldspar), muscovite mica and quartz, the main igneous rock-forming minerals can be divided into two groups. These are, firstly, the group of individually discrete ferromagnesian minerals, where the silica tetrahedra are joined by Fe^{++} and Mg^{++} ions (and in some instances certain others), and, secondly, the continuous series of plagioclase feldspars in which varying proportions of the Si^{++++} ions of the silica tetrahedra have been replaced by Al^{+++} ions, plus additional cations to compensate for the resulting loss of positive charge. The weakest of the common ferromagnesian minerals is olivine (($MgFe)_2SiO_4$), composed of isolated silica tetrahedra linked on all sides by Mg^{++} and Fe^{++} ions, giving the silica : oxygen proportion of 1 : 4. Augite ($Ca(Mg, Fe, Al)Si_2O_6$) is rather more resistant, in that the silica tetrahedra form single chains by sharing one oxygen atom (silica : oxygen ratio

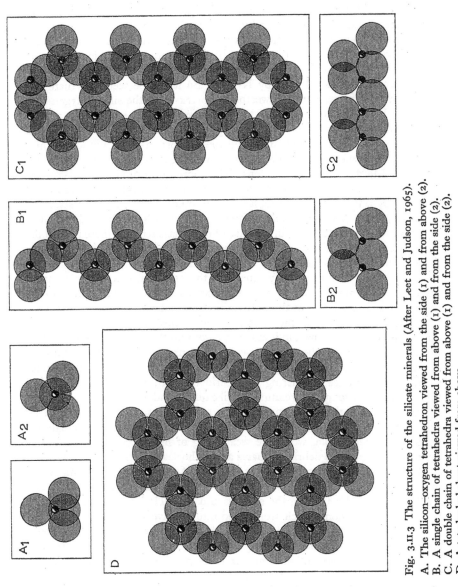

Fig. 3.11.3 The structure of the silicate minerals (After Leet and Judson, 1965).
A. The silicon–oxygen tetrahedron viewed from the side (1) and from above (2).
B. A single chain of tetrahedra viewed from above (1) and from the side (2).
C. A double chain of tetrahedra viewed from above (1) and from the side (2).
D. A tetrahedral sheet viewed from above.

$= 1 : 3$) (fig. 3.II.3(b)), but are weakly linked on four sides by other cations. Hornblende (of complex formula, of the general form $(OH)CaMgFeAlSiO_n$) has double silica tetrahedron chains (fig. 3.II.3(c)) linked by other ions (silica : oxygen ratio $4 : 11$); and in biotite mica the tetrehedra form plates with each tetrahedron sharing three oxygen atoms with their neighbours (silica : oxygen ratio $= 2 : 5$) (fig. 3.II.3(d)), in the other dimension two silica sheets being joined in a sandwich by Al^{+++}, Fe^{++}, and Mg^{++} ions and these sandwiches very weakly linked by K^+ ions. So it can be seen that there is a general increase of resistance to weathering break-up as the silica : oxygen ratio increases, but even in the case of the biotite break-up is easy in the parallel planes containing the K^+ ions. With the plagioclase feldspars there is a similar sequence of resistance, depending upon the proportion of Si^{++++} ions, which have been replaced by Al^{+++} ions, together with the additional charges provided by Ca^{++}, Mg^{++}, Na^+, and K^+ ions. The more substitutions of Al^{+++}, the weaker the structure becomes, such that calcic plagioclase (anorthite; $Ca(Al_2Si_2O_8)$), in which Al^{+++} replaces every other Si^{++++}, is less resistant to weathering than sodic plagioclase (albite; $Na(AlSi_3O_8)$), where Al^{+++} only replaces every fourth Si^{++++}. In the case of orthoclase feldspar there is a similar replacement ($K(AlSi_3O_8)$), and muscovite mica is similar to biotite, except that there are more Al^{+++} ions in place of some of the Fe^{++} and Mg^{++} ions giving a more resistant structure. Finally, quartz is the most resistant of the common igneous rock-forming minerals, in that all oxygen atoms are shared in all directions, giving a very resistant three-dimensional structure.

At higher temperatures silica tetrahedra are linked together more easily by other ions, and Al^{+++} ions find it easier to enter into the silica tetrahedra, so that it is a general rule that the minerals which form first at highest temperatures in a cooling melt (e.g. olivine, augite, and anorthite) form rocks which are most susceptible to surface weathering (e.g. peridotite), whereas rocks composed of the relatively low-temperature minerals, like quartz and orthoclase, are relatively resistant to weathering (e.g. granite) (fig. 3.II.4). Of course, where the minerals composing the rock have very different susceptibilities granular disintegration occurs, as, for example, when the weathering of hornblende and biotite in a quartz diorite causes a debris of quartz and orthoclase crystals to be formed.

Most weathering depends upon the presence of water, and the decomposition of silicate minerals is mainly accomplished by hydrolysis, in which the H^+ ions displace the metal cations in the silicates and the OH^- ions combine with the latter to form solutions which are washed into the rivers and seas. H^+ ions are chemically active because their small size enables them to penetrate many crystal structures, because they carry a large electrical charge relative to their size, because they form other compounds by providing hydrogen bonds, and because they can readily recombine with OH^- in some minerals (e.g. hornblende) to form water. At high concentrations of H^+ ions silica (SiO_2) and alumina (Al_2O_3) are linked by the H^+ ions which have displaced the metal cations to form complex clay minerals. Thus the type, as well as the degree, of weathering depends very much upon the amount of water available.

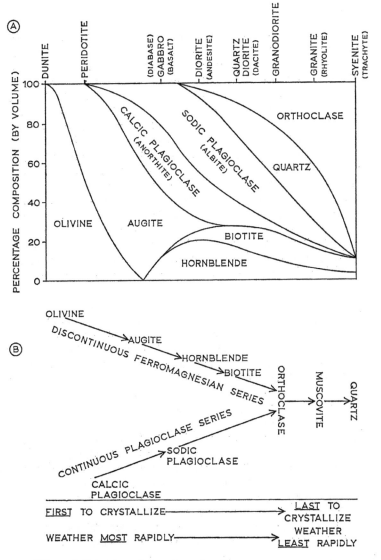

Fig. 3.11.4 The main igneous rock-forming minerals.
A. The average percentage composition of the most common igneous rocks.
B. The sequence of mineral crystallization and weathering.

There are a number of sources responsible for the production of H^+ ions in the water which weathers rocks:

1. The primary source is through the dissolving of CO_2 from the atmosphere, giving an average equilibrium pH of 5·7 at 25° C for rainfall and making it acid. The pH of rainfall, however, ranges from about 4 to more than 9 (for example, Hawaii 4·8–6·3, with a mean of 5·2; Uganda 5·7–9·8, mean 7·8; west coast of Ireland 5·9–7·6, mean 6·5), but it is difficult to generalize about variations on a global scale. Even in an area as small as Sweden, the rainfall is acid along the west coast and more neutral in the north. Apart from the natural decrease of dissolved chloride with distance from the coast, it has been suggested that areas of heavy rainfall of high intensity composed of large drops have lower acidity values because the rain has less opportunity for dissolving atmospheric CO_2. It seems that some tropical regions experience rainfall of much higher pH than that, for example, measured on the average in Northern Europe (5·47), due to some extent to the higher tropical temperatures which inhibit CO_2 solution.

2. The acidity of rain-water is largely irrelevant to many weathering processes, because it is usually changed drastically as soon as the water enters the soil.

    (a) CO_2 is dissolved much more readily from the soil atmosphere than from the free atmosphere.

    (b) Some clay minerals (e.g. bentonite) yield H^+ ions.

    (c) Acids are produced by humus and soil organisms. Controlled experiments with bacterial acids lasting only a few hours have shown that, although resisted by alumino-silicate minerals (indeed silica seems less soluble in humic acids than in pure water), muscovite and some basic minerals may lose $\frac{1}{4}$–$\frac{1}{2}$ of their weight by solution. However, the real effect of organic acids as a weathering process has not been fully investigated, although one suspects that it may be very great, especially in the humid tropics.

    (d) The roots of plants provide H^+ ions by osmosis, which they exchange for the nutrients provided by Ca^{++}, Mg^{++}, and, especially, K^+ cations. Some plants also accumulate significant quantities of dissolved silica, and 3·5% of the dry weight of some tropical hardwoods and up to 10% of bamboos is composed of silica. Assuming a conservative mean value of 2·5%, this would imply the removal of 0·4 tons of silica/acre/year in some tropical areas, which alone would account for 1 ft of denudation every 5,000 years on a basalt having 49% silica.

As weathering processes develop, however, by reactions involving the free H^+ ions, the acidity of water in soil and rock will decrease due to chemical reactions. In British Guiana, for example, rainfall of pH 7 quickly yields soil water of ph 8 on hornblende-rich rocks. Prolonged solution of quartz (pure silica) has little effect, maintaining a neutral pH of 6–7, but feldspars yield a solution of pH 8–10, augite 8–11, and hornblende of 10–11, again supporting the standard scale of chemical susceptibilities. Of course, different minerals are differently susceptible to chemical attack by soil-water solutions, as are different rocks lying in close juxtaposition, so that chemical weathering by altering the character of these fluids sets up chain reactions of weathering of a very complex kind.

1. Where the weathering fluid has a pH of 10 or more, Al_2O_3 is very soluble and SiO_2 relatively so (fig. 3.11.5), and both are carried away in solution, providing enough rainfall is available. Where rainfall is scanty, or evaporation is very high (or both), the decay products of Al_2O_3 and SiO_2 are not removed, but combine to form clay minerals such as montmorillonite and illite.

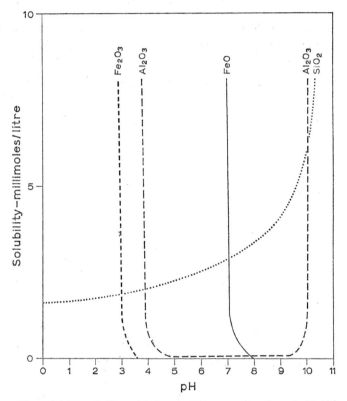

Fig. 3.11.5 The solubility of silica, alumina, and iron as a function of pH (After Keller 1957, and Loughnan, F. C., *Journal of Sedimentary Petrology*, Vol. 32, 1962).

2. Where the pH is more neutral (e.g. 7–8), as for example where heavy tropical rains are falling on a surface rock containing hornblende, Al_2O_3 is almost insoluble, whereas SiO_2 is still partly soluble (below this pH its solubility is very small, although it increases with temperature). Under these conditions hydrated Al_2O_3 remains as a residue, usually to form gibbsite ($Al_2O_3.3H_2O$), a valuable aluminium-ore mineral.

3. If soil water is acid, as for example in temperate latitudes, where rain falls on quartz-rich rocks covered in vegetation, the solubility of both Al_2O_3 and SiO_2 is very low, and both remain (providing the rainfall is not so unduly heavy as to flush away the H^+ ions) and combine to form clay minerals like kaolinite ($Al_2(OH)_2Si_4O_{10}$) together with quartz debris.

The weathering of a granodiorite under humid temperature conditions illustrates both the relative susceptibilities of its varied mineral constituents to weathering and the main types of products of igneous rock weathering:

1. Quartz (SiO_2). Very slight solution, but the main products are crystal fragments.
2. Orthoclase, combines with carbonic acid and water to produce soluble potassium carbonate (used by plants), the clay mineral kaolinite, plus some soluble silica and fine colloidal particles which are washed away.

$$2KAlSi_3O_8 + H_2CO_3 + nH_2O \longrightarrow K_2CO_3 + Al_2(OH)_2Si_4O_{10}.nH_2O + 2SiO_2$$

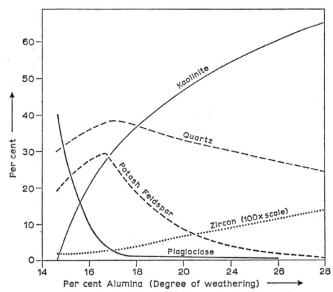

Fig. 3.II.6 Mineral-variation diagram of weathering of a granite gneiss under humid temperate conditions (After Goldich, S. S., *Journal of Geology*, Vol. 46, 1938).

3. Plagioclase. Anorthite and albite similarly combine with carbonic acid and water to produce soluble calcium and sodium bicarbonates, together with kaolinite.

$$CaAl_2Si_2O_8.2NaAlSi_3O_8 + 4H_2CO_3 + 2(nH_2O) \longrightarrow$$
$$Ca(HCO_3) + 2NaHCO_3 + 2Al_2(OH)_2Si_4O_{10}.nH_2O$$

4. Biotite combines with oxygen, carbonic acid, and water to produce soluble potassium and magnesium bicarbonates, limonitic iron, kaolinite, soluble silica, and water.

$$2KMg_2Fe(OH)_2AlSi_3O_{10} + O + 10\,H_2CO_3 + nH_2O \longrightarrow$$
$$2KHCO_3 + 4Mg(HCO_3)_2 + FeO_3.H_2O + Al_2(OH)_2Si_4O_{10}.nH_2O +$$
$$2SiO_2 + 5H_2O$$

5. Hornblende weathers similarly to biotite, but more rapidly.

Figure 3.11.6 shows the changes in relative mineral composition during the progressive weathering of a granite gneiss under humid conditions indicating the stability of quartz, as distinct from the feldspars, which decompose to produce the increasing clay content, with orthoclase being more resistant than plagioclase. The influence of climate on the mechanisms and products of igneous rock weathering is shown by recent work on a weathered quartz diorite in Antarctica, which, although apparently considerably decomposed, was found to be little changed chemically. No clay had been produced, the only chemical change being the oxidation of some iron, and the disintegration seemed mainly due to physical weathering by the growth of ice and sea-salt crystals. Where chemical weathering is dominant, however, igneous rocks break down to produce metal cations in solution, some silica in solution (much of colloidal size), a varying amount of silica wreckage (chiefly in the form of quartz particles), and clay minerals.

3. Weathering of sedimentary rocks

The breakdown of minerals forming igneous rocks thus leads to the production and isolation of more stable minerals, like clay and silica debris, as well as cations, largely of calcium. After the clastic debris is washed away by rivers the clay fraction of it is commonly separated from the coarser material, usually in a marine environment, to form beds which may be lithified to form *shale*. The composition of the shale is largely dependent on the types of clay minerals involved and partly on the environment of deposition. Their future weathering history is very much governed by the presence of metal cations in the open crystal lattice which controls their oxidation and hydration. The coarser clastic quartz material tends to be separated from the clay by the fluvial and marine agents of transportation and forms quartz sandstone (*orthoquartzite*). Obviously the quartz is very resistent to further weathering, and the weathering of sandstone is largely controlled by the material cementing the quartz grains. This is commonly iron, silica introduced subsequently in solution, or calcite and gypsum. The first is the most widespread; where the second occurs the resulting sandstone is extremely resistant to sub-aerial processes (as in the sandstone ridges of the Appalachians); and the third is usually the result of sub-aerial deposition in an arid environment. Formations like the Wingate and Entrada Sandstones of the Colorado Plateau are aeolian sands poorly cemented with calcite. A third type of clastic rock is *greywacke*, a poorly sorted sandstone containing at least 20% shale and characteristic of rapid sedimentation in geosynclines.

TABLE 3.11.1 Composition (% by weight)

Formula	Earth's crust	Average igneous rock	Average granite	Average basalt	Average sandstone	Average greywacke	Average arkose	Average shale	Average limestone
SiO_2	59·08	59·12	70·8	49·9	79·7	65·8	76·1	62·2	5·2
Al_2O_3	15·23	15·34	14·6	16·0	4·8	14·4	11·5	16·5	0·8
CaO	5·10	5·08	2·0	9·1	5·6	3·6	1·6	3·3	43·0
Na_2O	3·71	3·84	3·5	3·2	0·5	3·5	2·0	1·4	0·1
FeO	3·72	3·80	1·8	6·5	0·3	4·3	—	2·6	—
MgO	3·45	3·49	0·9	6·3	1·2	3·0	0·1	2·6	8·0
K_2O	3·11	3·13	4·1	1·5	1·3	2·1	5·7	3·5	0·3
Fe_2O_3	3·10	3·08	1·6	5·4	1·1	1·0	2·4	4·3	0·5
CO_2	—	—	—	—	5·1	1·6	0·4	2·7	41·9

The weathering of greywacke is primarily a function of the breakdown of the shale. The last important clastic rock is the least common – *arkose*. This is a sandstone containing less than 25% quartz and less than 20% shale, the rest being composed of feldspar. Although arkose is formed in a wide variety of environments from the breakdown of granitic rocks, it is most characteristic of rapid burial of partly weathered granite residue under arid conditions. Fossil alluvial fans flanking granite fault blocks have formed much of the present arkose, which is naturally quite susceptible to the further weathering of the feldspar if exposed at the surface. Of all the rocks exposed at the surface of the present continents, fully 52% are clays and shales and 15% the various types of sandstones. The metal cations, particularly Ca^{++}, which forms about $\frac{1}{6}$ of the total dissolved load of the rivers of the world, are washed into the oceans. Although the oceans are not saturated with these cations, the calcium is combined with CO_3 (dissolved from the atmosphere) and precipitated as reef and pelagic limestone by marine organisms. Limestone covers about 7% of the present continental surfaces, being mostly coral limestone, coral breccia, or pelagic limestone. Table 3.II.1 shows the major chemical compounds in some common rocks and indicates the possibility that most of the sedimentary rocks have been formed from the breakdown of igneous rocks.

Solution is, of course, an important weathering process for all rocks, but it is especially destructive to the carbonate sedimentary rocks. The breakdown of the two important carbonate minerals of calcite and dolomite under the action of carbonation by carbonic acid is the best example, and one which is most active at low temperatures:

$$CaCO_3 + H_2O + CO_2 \rightleftharpoons Ca(HCO_3)_2$$

$$\text{(Calcite)} \qquad \text{(Soluble calcium bicarbonate)}$$

$$CaMg(CO_3)_2 + 2H_2O + 2CO_2 \rightleftharpoons Ca(HCO_3)_2 + Mg(HCO_3)_2$$

$$\text{(Dolomite)} \qquad \text{(Soluble calcium and magnesium bicarbonate)}$$

The susceptibility of limestone to weathering partly depends upon its purity, but even more on the absolute amount of water available. Thus one finds that in desert areas limestone outcrops form residual hills and escarpments, whereas other rocks are more rapidly denuded. An example of the relative susceptibility of an arkosic sandstone, a granite, and a quartzite is given in fig. 3.II.7 relating to part of the Sangre de Cristo mountains in New Mexico.

Oxidation, the combination of oxygen with another atom causing the latter to lose an electron and take on a positive charge, is also a weathering process to which a wide variety of rock-forming minerals are susceptible. The simplest example is the production of the iron oxide hematite:

$$2Fe + 3O_2 \longrightarrow Fe_2O_3$$

Oxidation is probably almost entirely effected by the intermediary action of water, particularly in the soil, where there is a plentiful supply of CO_2 to the

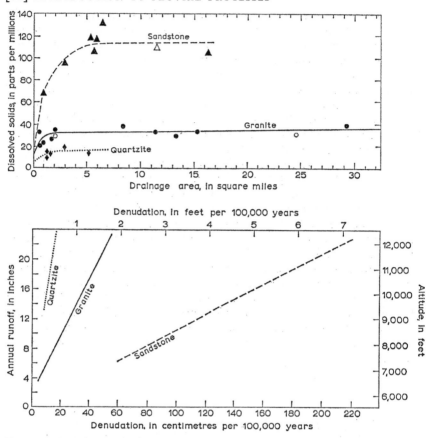

Fig. 3.11.7 Solution in the Sangre de Cristo Range, New Mexico (After Miller, J. P., *U.S. Geological Survey Water Supply Paper* 1535–F, 1961).

Above: Relation of dissolved solids in streams to drainage area (open symbols represent weighted mean values below tributary junctions). Each curve intersects the *Y*-axis at the average value of dissolved solids in snow (i.e. approx. 5 p.p.m.).
Below: Denudation as a function of altitude, calculated from the annual runoff, showing the relative resistance of quartzite, granite, and sandstone.

soil atmosphere by organic decomposition. In terms of sedimentary rock weathering, oxidation is particularly important in the weathering of the clays with especially open crystal lattices (e.g. montmorillonite), where oxygen combines with the Mg^{++} and Fe^{+++} ions.

One of the chief means of breaking down some clays, however, is by hydration – the simple adsorption of water into the crystal lattice with no fundamental chemical change, but accompanied by considerable swelling and the setting up of physical stresses. Where continuous cycles of wetting and drying occur, some clays become very fragmented. Clays are hydrous silicates containing metal cations. Their fundamental building blocks are sheets of silica tetrahedra sharing oxygen atoms with an associated octahedral sheet composed of O^{--} and OH^- ions, grouped around the metal cations (fig. 3.11.8). Kaolin possesses pairs of these sheets (fig. 3.11.8) closely bonded together with the small H^+ ions. This small space allows little further reaction and little ionic exchange to take place with the metal cations. Kaolin is produced by the alteration of feldspar-rich rocks under acid and humid conditions where most of the cations (except

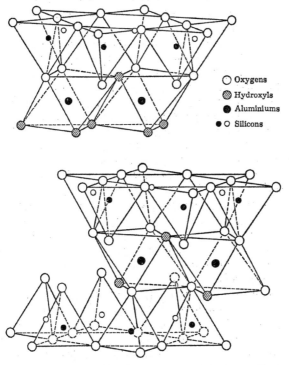

Fig. 3.11.8 The atomic structure of (*above*) kaolinite and (*below*) montmorillonite (After Grim).

Al^{+++}) are leached away and H^+ ions introduced. Illite is composed of octahedral sheets sandwiched between two tetrahedral sheets and the sandwiches strongly held together with K^+ ions. Although some internal chemical reactions occur, there is little ionic exchange or hydration. Illite derives mainly from the weathering of feldspars and micas under alkaline conditions with abundant Al^{+++} and K^+ ions. The most complex of the three common clay minerals is montmorillonite $((OH)_2(SiAl)_4(Al,Fe,Mg)_{2-3}(Na,Ca)_{1-3}O_{10})$ (fig. 3.11.8). Its structure is similar to illite, but the sandwiches are more widely spaced, allowing the entry of water and great expansion of the lattice (i.e. swelling). This allows much ionic exchange within the octahedral sheets, where Fe^{+++} and Mg^{++} are substituted for Al^{+++}, but the bonding remains very loose because the electrical charges between the ions never balance. Montmorillonite forms from basic rocks under alkaline conditions in the presence of Ca^{++} and Mg^{++} ions and a deficiency of K^+ ions. Besides influencing weathering, hydration affects the rates of erosion of shale and clay outcrops, for example, the sodium-rich parts of the Mancos Shale in the Colorado Plateau may, when subjected to free swelling, increase in volume by almost 60% under hydration, and repeated wetting and drying produces a surface layer of debris very susceptible to creep.

Another class of mechanical stresses in rocks due to the presence of water involves the growing of crystals in pore spaces and interstices. The effectiveness of ice-crystal growth has already been mentioned, but under certain conditions the crystallization of salts (sodium chloride, gypsum, calcite, etc.) is important in rock disintegration, granulation, and cavernous weathering. Salt crystallization exhibits preferred orientations and growing stresses similar to ice, and with 1% supersaturation calcite may crystallize against a pressure of 10 atmospheres – of the same order as the tensile strength of rocks. However, ice crystallization is different, in that all the solution enters into the solid phase and crystallization begins from the outside and proceeds inwards – both attributes allowing ice to develop higher growing stresses than salts. Crystallization occurs near the surface of an outcrop, where the rock is porous, where both salts and water are abundant, and where evaporation permits crystal growth. This type of weathering occurs in many environments, but is particularly effective in arctic and desert areas. In high latitudes the nuclei of snowflakes provide salt which, because melting and runoff is small, tends to accumulate near rock surfaces and disintegrate them. In arid regions the excessive evaporation causes salts to be drawn up from depth in capillary films and crystallization to occur at the surface, producing weathering which is particularly effective in shady locations. In general, the present desert areas have tended to be arid in the past, and therefore many of the underlying rocks are sources of salts which migrate to the surface.

It has often been assumed in the past that much desert weathering can be explained without recourse to the effects of water. However, despite the fact that diurnal temperature variations are often great, and that rocks are generally poor thermal conductors and composed usually of minerals having different coefficients of thermal expansion, observation and experiment have shown that thermal dilation is only important in breaking up rock surfaces in the presence of water. The existence of pronounced chemical rotting, particularly in shady sites where more moisture is available, shows that chemical weathering is dominant even in arid areas, as does the expansion of exfoliation shells on boulders, which is obviously due to expansion accompanying chemical alteration. Occasional rainstorms and, particularly, nocturnal dew form a significant supply of desert moisture, the presence of which can be detected several feet below the surface of boulders. Observations on a quartz monzonite boulder in the Mojave Desert have shown that a diurnal temperature range of 24° C would cause a significant linear expansion of 0·0084%, but that the temperature gradients within the rock are sufficiently uniform to allow the whole mass to expand and contract with little differential stress within it. The products of desert weathering, however, differ from those of more humid areas, being on the average rather coarser and not possessing such a high proportion of clay or organic material. These characteristics partly explain differences between arid and humid geomorphic processes and forms, in that creep of the organic- and clay-rich humid soils often contrasts with sheet erosion in desert areas, and that repose slopes in humid areas tend to be of lower angle.

Although in this discussion of the weathering of sedimentary rocks processes and environments have been mentioned which also relate to other rock types, the dominant primary weathering process of igneous rocks is hydrolysis, whereas solution, oxidation, and hydration are more important in most sedimentary rock decomposition. Metamorphic rocks are generally very susceptible to chemical weathering in the presence of water because the usual effects of metamorphism are to produce secondary high-temperature minerals of low stability, together with banding and schistosity, which encourages the entry of surface water. One major exception is recrystallized orthoquartzite, which is probably the most resistant of all rocks to weathering and universally forms strong relief.

4. The weathered mantle

Under suitable conditions, particularly in the humid tropics, weathering can proceed to considerable depths. Shales in Brazil have been altered down to 400 ft, and in Georgia granite has weathered *in situ* to 100 ft. Depth of weathering is controlled by the bedrock, climate, biological action, topography, and time, but even within a single rock body it can be extremely varied, especially where closely spaced joints expose a huge internal weathering surface (fig. 3.II.9), allowing weathering to proceed rapidly to considerable depths. Granite in Hong Kong is weathered to a maximum of 300 ft, but the removal of this mantle would expose a surface of considerable relief. The formation of tors and even tropical inselbergs has been ascribed to deep weathering around more-resistant and less-jointed parts of the rock body, accompanied or succeeded by the evacuation of debris either associated with rejuvenated stream downcutting or climatic change.

The mantle/rock contact exhibits considerable variety. Some rocks, notably basic rocks in the humid tropics, have a sharp transition a few mm wide between the weathered mantle and the bedrock, along what has been termed the 'weathering front' by analogy with the advance of some metamorphic processes. (Indeed, some authors refer to weathering as 'katamorphism'.) Usually, however, the contact is gradational, and granites in the tropics commonly have a transitional layer several metres thick. It appears to be sharpest where the rock is least permeable, where a water-table lies close to the surface, where the minerals weather rapidly, so that the rock weathers uniformly layer by layer from the top down, or where there are few minerals resistant to weathering. Much of the confusion of the contact in granite is due to the high proportion and irregular distribution of stable quartz. It should be stressed that erosion can produce a sharp topographic surface even with a transitional weathering contact, because relatively little weathering is necessary to loosen bedrock sufficiently for erosion to take place.

Rate and character of weathering is thus affected by the texture, fabric, and structure of rock bodies, as well as by their mineral composition. Even where solution is dominant in limestones, thin black rendzina soils remain under humid conditions, supporting short grass. The biological influence is important, not

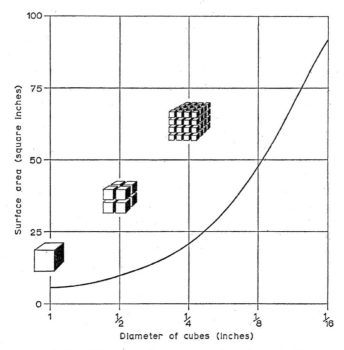

Fig. 3.II.9 The relation of the exposed surface area of a cube to the diameter of smaller cubes into which it is decomposed (Adapted from Leet and Judson, 1965).

only because humic and bacterial acids assist in mineral decomposition and because roots remove K^+ ions but because the accumulation of humus in the upper (A_0) layer of the soil (fig. 3.II.10(a) and (d)) both improves aeration and increases infiltration and water-holding properties. Indeed, much of the climatic influence over soil production is effected by the *soil climate*, in particular by the amount and movement of infiltrating water. Under humid conditions (e.g. pedalfer soils) the amount of clay is greater (fig. 3.II.10(c)) and the leaching of the silica-enriched A horizon is accompanied by deposition of clays, colloids, and cations in the Al^{+++} and Fe^{++} enriched B horizon. Under dryer conditions (e.g. pedocal soils) the discontinuous downward transport by water means that horizons are less well developed, and the periodic upward movement of alkaline ground water usually causes calcium carbonate (fig. 3.II.10(b)) and salts to accu-

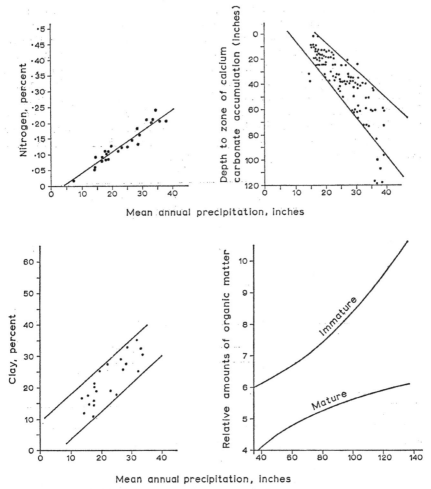

Fig. 3.11.10 The relation of mean annual precipitation to:

Top left. Nitrogen content.
Top right. Depth to zone of calcium carbonate accumulation (Data from Jenny and Leonard, 1934, and Russell and Engle, 1925).
Bottom left. Percentage of clay (Data from Jenny, 1941).
Bottom right. Relative amounts of organic matter
(Mostly from Leopold, Wolman, and Miller, 1964).

mulate below the surface which precipitation is insufficient to leach away. Topography represents another class of factors influencing weathering and soil formation, partly because of its control over micro-climate, vegetation, and drainage but also because downslope movements of soil influence soil thickness and structure.

Absolute rates of weathering are less easy to estimate than relative rates. What estimates have been made rely on such datable events as volcanic eruptions, Pleistocene stages, and archaeological structures. Absolute rates are extremely varied, depending on rock type and climate, and Kellogg was of the opinion that an inch of soil could be formed in any time between 10 minutes and 10 million years! Indeed, 14 in. of crude soil were formed on Krakatoa pumice in 45 years, 12 in. of soil developed on the top of the calcareous slabs forming the walls of Kamenetz Fortress in the Ukraine in 230 years, and 1·8 m of clayey B horizon were developed within pyroclastic material on the island of St Vincent in some 4,000 years. On the other hand, many glacially-scoured rock surfaces show little evidence of 10,000 years of post-Glacial weathering.

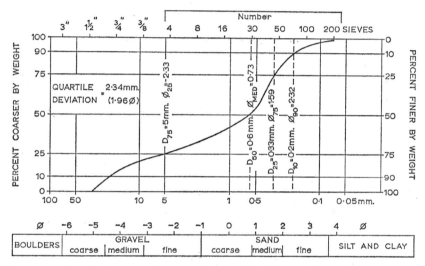

Fig. 3.II.11 A grain-size curve for a poorly-sorted glacial outwash.

From the practical point of view, the most important aspects of rock weathering are the chemical and physical characteristics of the weathered material. Some features of the chemical composition have already been mentioned and will be referred to again elsewhere in this volume. The simple parameters of the grain-size characteristics are shown in fig. 3.II.11. The two most important classes of size parameters are measures of absolute size and the range of sizes present, not least because they strongly influence permeability, shearing resistance, stability, and frost-heaving characteristics of the weathered material. These parameters are expressed in terms of diameter of particle (in mm) on a logarithmic scale, because of the huge range of sizes present in many soils which contain both gravel sizes and minute clay particles. Because of this range it has become the practice to use the ϕ (phi) scale, where $\phi = -\log_2$ diameter (mm), such that 1 mm = 0 ϕ, 2 mm = −1 ϕ, 4 mm = −2 ϕ, $\frac{1}{2}$ mm = 1 ϕ, $\frac{1}{4}$ mm = 2 ϕ, etc. Because the engineering properties of soils are so much influenced by the proportions of finer material present, another statement of size is the D scale; D being the diameter (in mm) for which certain percentages by weight are finer. Thus the median size (ϕ_{med}) is D_{50}, and the very diagnostic size for which 10% is finer (ϕ_{90}) is D_{10}. One of the simplest measures of range of sizes present in a soil is the quartile deviation equal to $\dfrac{\phi_{75} - \phi_{25}}{2}$.

REFERENCES

BECKINSALE, R. P. [1966], Soils: Their formation and distribution; Chapter 24 in *Land, Air and Ocean*, 4th edn. (Duckworth, London), pp. 361–99.

BURMISTER, D. W. [1951], *Soil Mechanics*; Vol. 1 (Columbia University Press, New York), 155 p.

DAVIS, K. S. and DAY, J. A. [1964], *Water: The Mirror of Science*; The Science Study Series Number 21 (Heinemann, London), 195 p.

GILLULY, J., WATERS, A. C. and WOODFORD, A. O. [1960], *Principles of Geology*; 2nd ed. (Freeman, San Francisco), 534 p. (especially pages 43–57).

HENDRICKS, S. B. [1955], Necessary, Convenient, Commonplace; in *Water*, U.S. Department of Agriculture Yearbook (Washington, D.C.), pp. 9–14.

KELLER, W. D. [1957], *The Principles of Chemical Weathering* (Lucas Bros., Columbia, Missouri), 111 p.

KELLER, W. D. [1957], *Chemistry in Introductory Geology* (Lucas Bros., Columbia, Missouri), 84 p.

KRUMBEIN, W. C. and PETTIJOHN, F. J. [1938], *Manual of Sedimentary Petrography* (Appleton-Century-Crofts Inc., New York), 549 p.

LEET, L. D. and JUDSON, S. [1965], *Physical Geology*; 3rd edn. (Prentice-Hall, New Jersey), 406 p. (especially pp. 76–90).

LEOPOLD, L. B., WOLMAN, M. G. and MILLER, J. P. [1964], *Fluvial Processes in Geomorphology* (Freeman, San Francisco), 522 p. (especially pp. 97–130).

MOHR, E. C. J. and VAN BAREN, F. A. [1954], *Tropical Soils*; (Interscience, London and New York), 496 p. (especially pp. 133–78).

REICHE, P. [1950], A Survey of Weathering Processes and Products; Revised Edn., *University of New Mexico Publications in Geology*, No. 3, 95 p.

4.II. Soil Moisture

M. A. CARSON

Department of Geography, McGill University

A large literature exists on the physics of soil moisture and upon the way in which soil water influences the nature of the soil in which it exists. This essay is not intended as a summary of these topics. Such standard works as *Soil Physics*, by Baver [1956], and *Soil*, by Jacks [1954], already serve this purpose admirably. The purpose of this essay is to outline the part played by soil moisture in some aspects of the denudation of the landscape.

1. The nature of soil moisture

A number of forces are capable of attracting water into dry soil. One is the simple affinity of the soil particles for water vapour in the soil atmosphere, although such hygroscopic water forms only a very small percentage of the water existing in most soils. A more important mechanism is the capillary suction which exists on the menisci of water films in contact with soil particles. This suction is usually explained by analogy with the rise of water in a capillary tube. Insertion of a thin

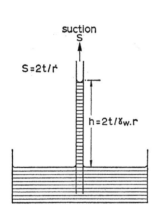

t : surface tension of water

δ_w: density of water

r : radius of tube and meniscus

Fig. 4.II.1 Suction on water in a capillary tube.

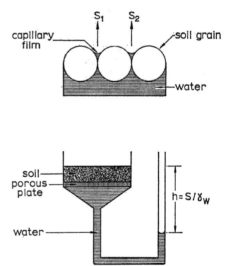

Fig. 4.II.2 Suction on water in soil.

tube into a tank of standing water (fig. 4.II.1) produces a rise in the level of the water in the tube relative to the level outside it. The extra height of the water in the tube is attributable to capillary suction acting against the force of gravity, which, alone, would maintain the same level inside and outside the tube. The magnitude of the suction is determined by the surface tension of the water and the radius of the meniscus. The height of capillary rise in the tube is determined by the ratio of the suction and the weight of a unit volume of water. The menisci in the pores of soil possess similar suction (fig. 4.II.2), and this is also measured by noting the amount of water displaced against gravity. As capillary water is

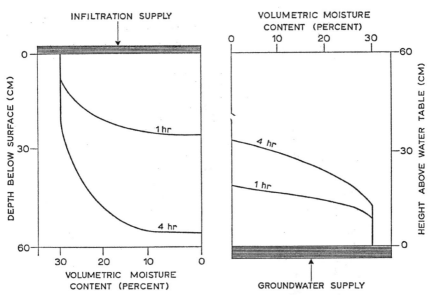

Fig. 4.II.3 Distribution of water in a sandy soil mass during (*left*) surface infiltration, and (*right*) capillary rise from a water table (After Liakopoulus, 1965a).

drawn out of a soil mass, the water which remains in the soil occupies smaller and smaller pores, and the suction on this water increases in precisely the same way as the height of rise in a capillary tube increases with decreasing radius of the tube. The capillary suction in a soil, together with the gravity force, determines the major movement and distribution of water in soil.

The pattern of change in the distribution of water in a soil mass during infiltration from water on the surface, and also during the entry of water upward from a ground-water system, has been treated by many workers. The similarity between the two processes has been emphasized by Liakopoulos [1965a] and is demonstrated in fig. 4.II.3. The entry of water into a soil mass from the surface proceeds by the downward advance of a wetting front where there is a sudden change from wet to dry soil. The amount of moisture in the soil shows little change with depth at a distance behind the wetting front, although it decreases rapidly in the area immediately behind it. A similar pattern of change occurs with the upward advance of a wetting front from a ground-water supply. In the soil just beneath the wetting line there is an increase in water content with depth until a constant moisture content is attained. This special value is identical in the two cases, and in the sandy soil used by Liakopoulos was about 30% by volume, which represents about three-quarters saturation of the pore space. These results agree with those of Bodman and Colman [1943] and other early workers.

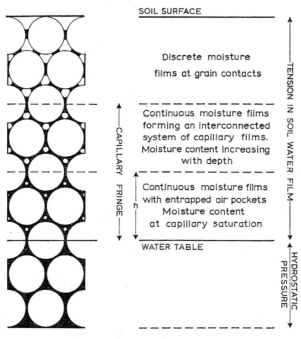

h=height of equivalent rise in a capillary tube

Fig. 4.11.4 The zones of capillary moisture in soil.

The ultimate equilibrium position to which the system tends differs in the two cases. In the case of capillary rise (fig. 4.11.4) a number of distinct zones exist. Immediately above the water-table is a belt of soil with constant moisture content at capillary saturation. Above this there is a systematic decrease in the amount of moisture in the soil. The area of continuous capillary water above the water-table is termed the capillary fringe, and above it the moisture films separate into discrete menisci. The thickness of the capillary fringe depends upon the size of the pores in the soil mass in much the same way as the height of water in a capillary tube depends upon the radius of the menicus. The theoretical extent of the capillary fringe is very great in the case of a clay with very small pores, whereas it is negligible in a sandy soil. The rate of capillary flow in a soil with very small pores is usually so slow, however, that the state of ultimate equilibrium is never attained, although, even then, the extent of this zone is appreciably greater than in sandy soils.

The ultimate moisture distribution in the infiltration case must differ from the capillary rise case, since the water supply will eventually exhaust itself. The pattern of moisture redistribution after the stage when all the water has entered the soil mass is unfortunately not clearly known. All the water above the wetting front will not continue to drain down through the soil. The gravity force will draw some water through the initial wetting front, but the development of menisci and capillary suction in the soil mass will tend to oppose this force. This

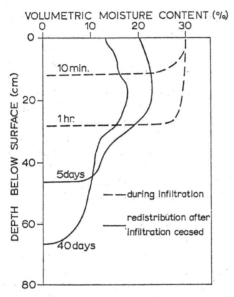

Fig. 4.11.5 Distribution of water in a sandy soil mass during and after infiltration from the surface (After Liakopoulos, 1965b).

was noted by early workers who distinguished between 'movable' and 'immovable' water. A soil which has just shed all its drainable water and still retains its maximum capacity of water which is immovable under the attraction of gravity alone was stated to exist at its *field capacity*. In this state the soil water would be held by a suction of about 20–40 in. of water. The amount of moisture in the soil at field capacity would depend upon the number of small pores and was thought to range from about 30% by volume for a clay soil to less than 10% for sandy soils. The practical value of the concept of field capacity has been criticized by a number of workers. The tests by Liakopoulos [1965b] show that even in a fine sand (fig. 4.11.5) drainage past the initial wetting front proceeds a long time after infiltration has ended.

The water which drains past the initial line of the wetting-front moves towards an underlying ground-water system. During this movement the water is still subject to lateral capillary suction and, as pointed out by Sherman [1944], it may never reach the water-table unless most of the soil pores are large. This depends also on the height of the water-table. The subsequent history of movable water beneath the initial wetting front depends, in addition, on the nature of the solid rock underlying the soil mantle. A shattered mass of rock with large gaps will facilitate downward percolation to the water-table, whereas a highly impermeable stratum may lead to a temporary ground-water system above the main one at depth.

The picture presented above is a simplified account of the movement and distribution of water in a soil mass during infiltration from the surface on level ground. A very different situation must exist on a steeply sloping land surface. The tests by Whipkey [1965] suggest that a wetting front advances into the soil mantle in much the same way as on a level surface. These tests do suggest, however, that the soil may attain complete saturation. Another feature is that during and after infiltration there is a marked downslope flow within the soil mass, and this is especially marked where an impermeable soil layer retards vertical entry of water. Such water will by-pass the underlying ground-water system and return to the stream through the soil mantle. It is important to realize therefore that the infiltration of storm water into a soil may differ radically on slopes from a flat land surface. These tests indicate that temporary water systems perched above the main water-table may occur often during prolonged rainstorms and offer a mode of subsurface flow entirely different from the pattern which occurs within a soil mass beneath flat ground.

2. Soil moisture and denudation

The most obvious part played by soil moisture in the denudation of the land-scape is the assistance given in the weathering of solid rock masses into loose debris and the direct transport of material as solutes and colloids out of the waste mantle. This direct loss of material from the waste mantle by moving water has attracted little interest in the past. Attention has been focused upon the conditions which induce subsequent redeposition at another level within the mantle and thus create a soil profile.

There is never uniformity with depth in a soil mantle. The upper parts of the mantle are more humic, and the lower parts tend to be more moist. Soil minerals which are unstable in the upper levels and taken into solution by the soil water may attain greater stability at depth in the mantle and redeposition occurs. This eluviation and illuviation inevitably accentuate the differences between the upper and lower parts of the mantle, and a soil profile ensues. A well-leached soil commonly shows a pallid layer of silt and sand grains which overlies a horizon of illuviated clay and other minerals. The emergence of a soil profile is complicated in dry areas by the upward movement of soil water and the deposition of salts in the upper crust with evaporation.

The classic issue of the development of soil profiles usually assumes a flat surface. The loose material which mantles most hillslopes often shows, in contrast, little sign of illuviation: the only noticeable change with depth is an increase in the amount of unweathered debris. The absence of a soil profile on slopes may be due to many reasons. The soil mantle on a hillslope is subject to continual erosion of different types, and it is maintained only by the compensating supply of new soil through the weathering of the underlying solid rock. Although a soil mantle may exist permanently on a slope, it exists in a dynamic state (Nikiforoff, 1949), and a particular mass of soil may not stay on the slope for a sufficiently great length of time to develop a profile. Another explanation may be the tendency of soil water to seep downslope rather than vertically, as noted by Whipkey [1965], and under these circumstances a downslope change in soil type rather than a vertical profile might materialize. Such a pattern would

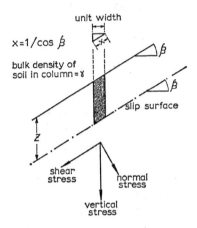

vertical stress = $\gamma . \bar{z} / x = \gamma . \bar{z} . \cos \beta$

shear stress = $\gamma . \bar{z} . \sin \beta / x = \gamma . \bar{z} . \sin \beta . \cos \beta$

normal stress = $\gamma . \bar{z} . \cos \beta / x = \gamma . \bar{z} . \cos^2 \beta$

Fig. 4.11.6 The stresses imposed by gravity at a point in a soil mass.

be comparable to the catena sequence, except that it would be the product of subsurface soil water.

The transport of material out of the soil mantle by moving water within the soil mass is only a minor process, at least directly, in the denudation of the landscape. A residual soil mass will always survive this process, and much of the material lost from one part of the mantle may be expected to be deposited in another part. The major landforms are shaped by the processes of soil wash and mass-movement of different types which act upon this residual mantle. Soil moisture has a distinctive role in each of these. The influence of antecedent soil moisture in determining the amount of surface runoff and the extent of soil wash is discussed in Chapter 5.1. The less obvious role of soil moisture in the mass movements of shallow landslides and seasonal soil creep is discussed here.

Soil moisture plays a vital role in determining the stability of a soil mantle on a hillside. The shear stresses which exist within a soil mass and the shear strength to withstand these stresses both depend partly upon the amount of moisture in the soil. The shear stress on the plane of failure in fig. 4.11.6 is dependent upon the angle of the slope, the depth of the plane, and the density of the soil mass above it; the last of these will vary with the amount of moisture in the soil. The shear strength of a soil mass is derived from a cohesive element and a frictional component. The amount of internal friction developed along a plane of failure depends upon the effective stress normal to that plane as well as the angle of shearing resistance of the soil. The effective normal stress itself depends not only upon the component of the weight of the soil column at right angles to the

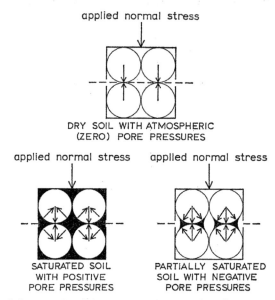

Fig. 4.11.7 The influence of moisture on pore pressures in soil.

plane (fig. 4.11.6) but also upon the pressure in the pores of the soil. This is shown in fig. 4.11.7. In a soil mass which is completely dry the normal stress applied by the overlying material is neither supplemented nor alleviated by the air pressure in the pores, since the pressure is atmospheric. When the soil pores are partly filled with water the pressure in the water films under the menisci is less than atmospheric (Skempton, 1960), so that the overall pore pressure is negative and a suction force augments the applied normal force in drawing the soil grains together. In a soil mass which is saturated with free-draining water the pore pressures are positive relative to the atmospheric datum, and this acts against the applied normal stress. The moisture content of the soil, through its influence on the effective normal stress and thus on the amount of internal friction which may be developed upon a potential plane of failure, is a vital consideration in the stability of a hillside soil mantle. It has indeed been suggested by Vargas and Pichler [1957] that the majority of natural landslides owe their origin to the development of positive pore pressures in a soil mass during prolonged rainstorms.

An implication of this is that, assuming that a soil mass is typified by particular pore-pressure values, the angle of limiting stability in any area will congregate around particular values determined by the pore pressure and the shear strength of the soil. The maximum angle of a stable slope occurs when the shear stress and the shear strength of the soil are just balanced. In the situation depicted in fig. 4.11.6 this is given by:

$$\gamma \cdot z \cdot \sin \beta \cdot \cos \beta = c' + (\gamma \cdot z \cdot \cos^2 \beta - u) \tan \phi'$$

where γ is the density of the soil mass;

z is the depth of the plane of failure;

β is the angle of slope;

c' is the cohesion of the soil;

u is the pore pressure;

ϕ' is the angle of shearing resistance of the soil.

The effect of pore pressure is most conveniently illustrated by dealing with soils which have negligible cohesion. The maximum stable slope in the situation in fig. 4.11.5 is then given by:

$$\tan \beta = \tan \phi' \left(1 - u/(\gamma \cdot z \cdot \cos^2 \beta)\right)$$

Soil mantles which never attain complete saturation and never fully dry out will always possess negative values of u, the pore pressure, and may thus stand at angles which exceed the angle of shearing resistance of the soil material. Schumm [1956] suggested that this occurs on 40–45-degree badland slopes in South Dakota, where there is sufficient silty material to provide lasting capillary suction. In contrast, soils which are essentially loose rock fragments are unlikely to attain complete saturation when they mantle hillsides due to the large pores, and for the same reason they are unlikely to maintain capillary water films permanently. They are thus characterized by pore pressures which are essentially atmospheric ($u = 0$), and the maximum stable slope is in this case the same as the angle of shearing resistance of the material. This is very probably the reason why so many scree slopes exist at angles near to 35 degrees, since this value approximates the angle of shearing resistance of small loose rocky matter. The majority of hillslopes, however, at least in humid areas, stand at angles which are less than the angle of shearing resistance of the soil mantle, and it seems very likely that this is due to the development of positive pore water pressures at times of prolonged rainstorms which give rise to perched water-tables. The pore-water pressures in free-draining water depend upon the pattern of flow, but in the case of ground-water flow parallel to the surface (fig. 4.11.8) the pressure at any point is given by:

$$u = \gamma_w \cdot z \cdot \cos^2 \beta$$

where γ_w is the density of water.

Substitution of this value in the previous equation gives:

$$\tan \beta = \tan \phi' \, (1 - \gamma_w/\gamma)$$

and since the bulk density of most surface soils is about twice the density of water, this indicates that the maximum stable slope under these circumstances should approximate to half the angle of shearing resistance in tangent form. The work of Skempton and DeLory [1957] suggests that, in the London Clay at least, this hypothesis is supported by the field evidence: angles of limiting slope approximate 8–9 degrees, and this agrees with a residual angle of shearing

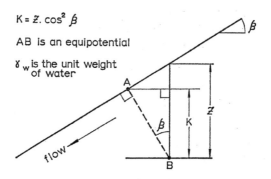

$K = Z \cdot \cos^2 \beta$

AB is an equipotential

γ_w is the unit weight of water

pore‑water pressure $= 0$ at A and u_1 at B

positional potential $= \gamma_w Z \cos^2 \beta$ at A and 0 at B

total head at B $=$ total head at A

$$\therefore u_1 = \gamma_w Z \cos^2 \beta$$

Fig. 4.11.8 The relation between pore-water pressure and depth in a soil mass with ground-water flow parallel to the surface.

resistance which is near to 16 degrees. In the case of an area which experiences widespread artesian pore-water pressures the limiting slope angle will be less than the value predicted on the basis of the previous model, although such pressures are possibly rather rare.

Soil moisture thus plays a conspicuous part in the denudation of the landscape under landslides and, through the attainment of particular values of pore pressures, may lead to the emergence of special angles of limiting slope in any one area. A hillside which is stable against rapid mass-wasting is not immune to still further transport of debris downslope, and in humid areas subsequent denudation is mostly through moisture-induced soil creep.

There are two types of soil creep which act upon hillslopes. One is the unidirectional movement downslope under the impetus directly of gravity and designated by Terzaghi [1950] as shear creep. The rate of this type of movement is probably very slow. Superimposed upon this continuous creep is seasonal soil creep. This is produced by random and cyclic disturbances which operate on flat land as well as on slopes. In the latter case, however, they are given a systematic bias downslope by the component of the gravity force. The major mechanism underlying movement in seasonal soil creep in humid temperate areas is the expansion and shrinkage of soils with changes in the moisture content of the soil. Some pioneer tests by Young [1960] indicated that under this mechanism alone soil creep might be expected to transport 0·5–1·0 cm^3 of soil past a line 1 cm wide across the slope in an average year. Subsequent work (Young, 1963) suggests that this approximates the actual amount of soil creep on slopes in the British Isles, and the conclusion that the bulk of seasonal soil creep is due to changes in the amount of moisture in the soil is supported by other workers. The creep rate clearly depends very much on the soil type on the slope. Sandy soils swell very little and clays considerably when water is absorbed, and it might be expected that the creep rate would increase with the amount of colloidal material in the soil. This is a possibility, although other features, such as angle of slope, may influence the rate of soil creep.

One of the major controversies in geomorphology is whether the processes of denudation cause hillslopes to decline in steepness or retreat at an unchanging angle. It is doubtful whether soil creep will produce either of these two changes on a straight hillside. The decline or the retreat of a straight hillslope demands that there is a systematic increase in the discharge of soil moved at different points downslope. This, in the case of soil creep, means that there must be one or both of an increase in the rate of soil creep and an increase in the thickness of the moving mantle downslope. The meagre evidence available suggests that neither of these occurs and, by implication, a straight slope acts essentially as a plane of transport for material from upslope with no effective erosion on the straight slope. The product of this is the replacement of the straight slope by the encroachment downslope of it of a convex hilltop, and there is neither retreat nor decline in the straight slope during this sequence.

The evolution of a landscape which is stable against rapid mass-wasting is, at least when it is free from the action of moving ice and the particular effects of karst limestone, determined primarily by the strength of soil wash as against soil creep. The balance between these two processes may depend a great deal on the soil type. Schumm [1956] showed this in the badlands of South Dakota and Nebraska: some clays absorbed all surface water and were subject only to creep while more impermeable clays induced surface run-off and soil wash. The effect of micro-climate on the balance between the two processes was discussed by Hack and Goodlett [1960]: moist slopes in the Appalachians appeared to be subject mostly to creep, while soil wash dominated on dry slopes. The major distinction between the two processes undoubtedly occurs with the differences in climate on the world scale. Soil creep in semi-arid areas is usually dwarfed in importance by the vast amount of soil wash which occurs in torrential storms on bare hillslopes. The results of work in humid areas suggests, in contrast, that soil creep is far more important and that it is only in the most intense and prolonged rainstorms that run-off and soil wash may take place. It is perhaps not mere coincidence, therefore, that the slopes of arid areas are dominated by straight profiles and sharp crests, while the upper hillslope convexity assumes its most distinctive development in the humid temperate areas of the earth.

The existence of soil moisture is clearly fundamental to the major processes of denudation. It is not surprising that differences in soil type and climate, affecting the amount and movement of soil moisture, are translated into differences in the efficacy of these processes and thus mirrored in the earth's landforms.

REFERENCES

BAVER, L. D. [1956], *Soil Physics*; 3rd Edn. (New York).

BODMAN, G. B. and COLMAN, E. A. [1943]. Moisture and energy conditions during downward entry of water into soils; *Proceedings of the Soil Science Society of America*, 8, 116–22.

HACK, J. T. and GOODLETT, J. C. [1960], Geomorphology and forest ecology of a mountain region in the central Appalachians; *U.S. Geological Survey Professional Paper* 347.

JACKS, G. V. [1954], *Soil* (Edinburgh), 221 p.

LIAKOPOULOS, A. C. [1965a], Theoretical solution of the unsteady unsaturated flow problem in soils; *Bulletin of the International Association of Scientific Hydrology*, 10, 5–39.

LIAKOPOULOS, A. C. [1965b], Retention and distribution of moisture in soils after infiltration has ceased; *Bulletin of the International Association of Scientific Hydrology*, 10, 58–69.

NIKIFOROFF, C. C. [1949], Weathering and soil evolution; *Soil Science*, 67, 219–30.

SCHUMM, S. A. [1956], The role of creep and rainwash on the retreat of badland slopes; *American Journal of Science*, 254, 693–706.

SHERMAN, L. K. [1944], Infiltration and the physics of soil moisture; *Trans. American Geophysical Union*, 25, 57–71.

SKEMPTON, A. W. [1960], Effective stress in soils, concrete and rocks; *Pore Pressure and Suction in Soils*, 4–16.

SKEMPTON, A. W. and DELORY, F. A. [1957] Stability of natural slopes in London Clay; *Proceedings of the 4th International Conference of Soil Mechanics*, 2, 378–81.

TERZAGHI, K. [1950], Mechanism of landslides; *Bulletin of the Geological Society of America*, Berkey volume, 83–122.

WHIPKEY, R. Z. [1965], Subsurface storm flow from forested slopes; *Bulletin of the International Association of Scientific Hydrology*, 10, 74–85.

VARGAS, M. and PICHLER, E. [1957], Residual soil and rock slides in Santos (Brazil); *Proceedings of the 4th International Conference of Soil Mechanics*, 2, 394–8.

YOUNG, A. [1960], Soil movement by denudational processes on slopes; *Nature*, 188, 120–2.

YOUNG, A. [1963], Soil movement on slopes; *Nature*, 200, 129–30.

5.I. Infiltration, Throughflow, and Overland Flow

M. J. KIRKBY

Department of Geography, Bristol University

When the rainfall that has not been intercepted by vegetation reaches the ground surface part of it fills small surface depressions (depression storage), part percolates into the soil, and the remainder, if any, flows over the surface as overland flow. Each component of this equation is highly variable, and depends not only on the intensity of the rainfall but also on soil, vegetation, and surface gradient. The amount of water intercepted by vegetation depends on the type of plants and their stage of growth, but it is usually close to a value given by all of the first 1·0 mm of rainfall and 20% of the subsequent rainfall in any one storm. Rainfall reaching the soil surface has to fill the small depressions on the surface before any overland flow can occur, even on a totally impermeable surface. This depression storage does not vary with the amount of rainfall but with the nature of the surface, especially with slope gradient, vegetation cover, and land-use practices. Under natural conditions depression storage absorbs about 2–5 mm of rainfall in any one storm. Contour ploughing is particularly effective in increasing depression storage by as much as ten times.

1. Infiltration

Infiltration rate is defined as the maximum rate at which water can penetrate into the soil. For initially moister soils the infiltration rate is lower throughout storms, and for all soils it decreases during the course of a storm. The rate at which water can travel through the soil depends on the number and size of pore spaces in the soil and the distribution of water within them. In effect, the infiltrating water has two components, a transmission component, which is constant and represents a steady flow through the soil; and a diffusion component, which is an initially rapid, and then an increasingly slow, filling-up of air-filled pore spaces, from the surface downwards. These components can be expressed in the infiltration equation (Philip, 1957):

$$f = A + B \cdot t^{-\frac{1}{2}} \tag{1}$$

where f is the instantaneous rate of infiltration;
$\quad$ t is time elapsed since the beginning of rainfall;
$\quad$ A is the 'transmission constant' of the soil; and
$\quad$ B is the 'diffusion constant' of the soil.

In this equation the transmission and diffusion terms can be identified with the two components of an idealized model. The transmission term represents unimpeded laminar flow through a continuous network of large pores. The diffusion term represents flow in very small discrete steps from one small pore space

TABLE 5.1.1 Variation of minimum infiltration rates with soil grain size, initial moisture content, and vegetation cover

(a) The effect of grain size in initially wet soils without vegetation cover

Grain size class	Infiltration rates (mm/hr)
Clays	0–4
Silts	2–8
Sands	3–12

(b) The influence of moisture content for Illinois clay-pan soils (after Musgrave and Holtan, 1964)

Initial moisture content (%)	Infiltration rates (mm/hr)		Poor weed cover
	Good grass cover		
	Topsoil > 13 in. thick	Topsoil < 13 in. thick	Topsoil < 13 in. thick
0–14	17	19	6
14–24	7	7	4
24 +	4	4	3

(c) The influence of ground cover for Cecil, Madison, and Durham soils (after Musgrave and Holtan, 1964)

Ground cover	Infiltration rate (mm/hr)
Old permanent pasture	57
Permanent pasture; moderately grazed	19
Permanent pasture; heavily grazed	13
Strip-cropped	10
Weeds or grain	9
Clean tilled	7
Bare ground crusted	6

to the next, in a random fashion. The only reason that a net diffusion flow results is that more pores are dry lower down, so that there is greater opportunity for downward movement than for upward. In an actual soil the two phases of this model cannot be separated, as all pores show a combination of the two types of behaviour, but equation (1) remains a good approximation for measured infiltration rates.

Table 5.1.1 shows some of the range of variation which can be expected in infiltration rates under a range of vegetation and moisture conditions, and in fig. 5.1.1 a comparison is made between typical infiltration rates and expected storm rainfall rates.

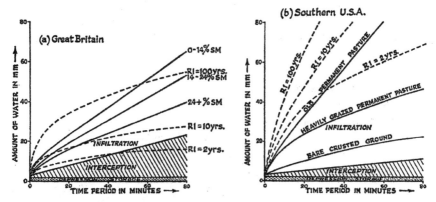

Fig. 5.1.1 Comparison of expected rainfall intensities with infiltration rate, interception, and depression storage, with representative values to demonstrate the relative frequency of overland flow in Great Britain and southern United States.

(a) Great Britain: rainfall (Data from Bilham, 1936), interception values for 100% vegetation cover, infiltration rates for Illinois clay-pan soils with good grass cover (Data from Musgrave and Holtan, 1964).

(b) Southern United States: rainfall (Data from Yarnell, 1935), interception values for 50% vegetation cover, infiltration rates for Cecil, Madison, and Durham soils (Data from Musgrave and Holtan, 1964).

2. Overland flow and throughflow

'Horton overland flow' is defined as overland flow which occurs when rainfall intensity is so great that not all the water can infiltrate, and is described by Horton [1945]. This type of overland flow is a fairly common phenomenon in semi-arid climatic conditions, but is relatively rare in humid and humid–temperate conditions. The role of vegetation is thought to be a critical cause of this distinction. Vegetation increases the infiltration rate by promoting a thicker soil cover, a better soil texture, and by breaking the impact of raindrops on the surface. Its effect on soil structure is mainly to build up an organic-rich A horizon with a relatively open pore structure and high permeability. If raindrops strike the surface without being impeded by vegetation fine material is thrown into suspension by the impact and is redeposited as an almost impermeable surface skin which can lower infiltration by as much as ten times. Vegetation therefore has a controlling influence on Horton overland flow by increasing both the initial depression storage and the infiltration rate, so that where a dense vegetation cover is established Horton overland flow is very unusual. Soil compaction by animals and vehicles reduces the infiltration rate while increasing depression storage, so that its net influence is problematic.

Within a small drainage basin, where soils are more or less homogeneous, it may be expected that interception, depression storage, and infiltration rate will not vary greatly, so that the conditions for Horton overland flow will be satisfied by comparable intensities and durations of rainfall, and overland flow will occur simultaneously all over the basin. Typical velocities for overland flow are about 200–300 m/hr, so that in a rainfall of 1 hour water from all points of a basin (200–300 m is a typical distance from divide to stream in Britain) will be reaching

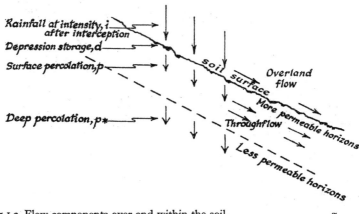

Fig. 5.1.2 Flow components over and within the soil.   z

a stream channel, and flow over the slope surface will have reached a steady state, represented by the equation:

$$q_0 = (i - f) \cdot a \qquad (2)$$

where q_0 is the overland flow discharge per unit contour length;
  i is the rainfall intensity after interception;
  f is the infiltration rate; and
  a is the area drained per unit contour length (equal to the distance from the divide if all the contours are straight lines).

Thus, provided that the rainfall intensity is high enough (or the infiltration rate low enough) for this type of overland flow the actual magnitude of the flow will be strongly dependent on the area or distance of overland flow, and will be almost independent of the storm duration, provided that it exceeds a reasonable minimum value. This is Horton's [1945] classic overland flow model.

Some of the water which infiltrates into the soil passes downward to recharge the water-table, and some, usually the greater part, flows down the hillside within the soil layers as 'throughflow' and ultimately contributes to streamflow (fig. 5.1.2). Within the soil, permeability varies and is generally highest in the open-textured organic A_0 horizon and the eluviated A_1 horizon. B horizons tend to be less permeable because clays are washed down into them, and in some cases because of the development of a hardpan. Conditions in the parent bedrock vary widely from limestones, with open solution fissures, to totally impermeable consolidated clays and shales. Wherever permeability is decreasing downward within the soil, and this occurs most commonly at the base of the A horizon, part of the water which is percolating downwards cannot penetrate into the lower layers fast enough, and is deflected laterally within the upper layer, as throughflow. This is similar to the production of Horton overland flow at the surface, except that within the soil the reduction of permeability is usually gradual, leading to a progressive deflection of throughflow. Table 5.1.2 shows an example of the progressive decline of permeability through the soil profile for a hillslope in Ohio, U.S.A.

TABLE 5.1.2 Variations of permeability and soil type with depth in an Ohio forest soil at a gradient of 15° (after Whipkey, 1965)

Soil depth (cm)	Textural class	Bulk density (gm/cc)	Saturated permeability (mm/hr)
0–56	Sandy loam	1·33	—
56–90	Sandy loam	1·41	286
90–120	Loam	1·78	17
120–150	Clay loam	1·80	2

If rainfall continues for a long time soil layers become saturated and throughflow is deflected closer and closer to the surface, so that the upper, more permeable soil layers are filling up from their bases because the throughflow is unable to carry away the water fast enough. In time the soil will become saturated right up to the surface and 'saturation overland flow' will occur. Under steady rainfall this condition will ultimately be attained under rainfall intensities much lower than are required to produce Horton overland flow. Since soil thickness and the velocity of throughflow vary much more over a small area than does permeability, and since the base of a slope tends to become saturated sooner than the divide, certain parts of a hillside are likely to produce saturation overland flow preferentially, in contrast to the rather widespread production of Horton overland flow, when it occurs at all.

Throughflow, travelling through soil pore spaces rather than over the ground surface, moves at very much lower velocities than overland flow. Rates of 20–30 cm/hr for throughflow are of the order of a thousand times lower than overland flow rates, so that periods of about 1,000 hours rainfall are needed for a steady state of flow to be achieved throughout an average basin. In practice, such a steady state is never attained for throughflow, and equation (2) for overland flow must be replaced with equation (3) for throughflow:

$$q_T = (p - f_*) \cdot v \cdot t \tag{3}$$

where q_T is the throughflow discharge per unit contour length;

p is the rate of surface percolation, equal to i or f, whichever is smaller;

f_* is the rate of infiltration at the base of the more permeable soil;

v is the velocity of throughflow; and

t is the time elapsed (strictly $v \cdot t$ should be replaced by the area/unit contour length within a distance $v \cdot t$ upslope).

In this equation the *time elapsed* is the most important control over the flow, in place of the *distance from the divide* for Horton overland flow. In reality, the flows are delayed by the time of transmission from the surface to a zone of decreasing permeability, a period which may be a matter of minutes or a few hours.

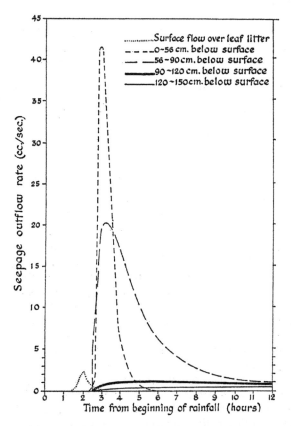

Fig. 5.1.3 Discharge hydrographs of flow within the soil resulting from a simulated storm of 5·1 cm/hr, lasting 2 hr, on a 16° slope which had previously drained for more than four days. The rapid, although small, Horton overland flow results from the initially low permeability of the dry surface soil, which rapidly increases with wetting. The lag before throughflow begins is the time taken for rain to infiltrate vertically to the 90-cm-deep, less-permeable interface (After Whipkey, 1965, p. 81).

If outflow from various depths in the soil is measured throughout a storm of uniform intensity there will be an initial transmission period of little or no flow, followed by a period in which the flow is increasing rapidly (though not linearly, because v in equation (3) is itself varying) until the moment when rainfall ceases, after which the flow will decrease more slowly than it increased. In other words, the flow should broadly resemble a flood hydrograph of a stream, despite the absence of surface run-off. In fig. 5.1.3 (Whipkey, 1965) actual values are shown for a rainfall intensity of 5·1 cm/hr falling for 2 hours on the soil whose properties are described in Table 5.1.2. It is apparent from fig. 5.1.3 that even at this high intensity Horton overland flow is negligible, and that throughflow from shallower soil layers is later than from deeper layers due to the time lag for transmission of water down through the soil, followed by saturation from the base up.

Equation (3) contains an unknown quantity, namely the velocity of through-flow, v. In order to evaluate the throughflow more exactly, Darcy's law, which states that flow through a permeable medium is proportional to the pressure gradient, must be combined with the continuity equation, which states that differences between inflow and outflow must be accommodated by changes in moisture content. For a soil layer of uniform permeability and moisture content (in depth) Darcy's law for soil on a slope is

$$Q = z \cdot \cos \alpha \left\{ K \cdot m \cdot \sin \alpha - D \cdot \frac{\partial m}{\partial x} \right\} \tag{4}$$

and the continuity equation is

$$\frac{\partial m}{\partial t} + \frac{\partial Q}{\partial x} = i \tag{5}$$

where Q is the downslope discharge measured in a horizontal direction;
$\quad$ x is the distance downslope measured in a horizontal direction;
$\quad$ K is the soil permeability (a function of soil moisture);
$\quad$ D is the soil diffusivity (a function of soil moisture);
$\quad$ α is the surface slope angle;
$\quad$ m is the soil moisture content;
$\quad$ z is the thickness of the soil layer;
$\quad$ i is the rainfall intensity after interception; and
$\quad$ t is the time elapsed.

Solution of these equations is necessarily numerical, but in a simple actual example of a uniform soil on a uniform gradient (Hewlett and Hibbert, 1966) estimates of permeability and diffusivity were as shown in Table 5.1.3; and calculated soil-moisture patterns during (a) uniform rainfall and (b) uniform drainage were as shown in fig. 5.1.4. It can be deduced from the data of Table 5.1.3 that the permeability (K) term in equation (4) is the more important for inter-mediate values of moisture, while the diffusivity (D) term becomes dominant at extreme low and high values of moisture. The low-moisture case is of little interest, but the high-moisture case, when the moisture content is approaching

TABLE 5.1.3 Values of permeability (*K*) and diffusivity (*D*) for varying soil moisture, computed from the results of an experiment carried out during soil drainage of a trough of soil inclined at 22° (Hewlett and Hibbert, 1963)

Moisture content (expressed as % of saturated moisture content)	Permeability (*K*) (cm/hr)	Diffusivity (*D*) (cm/hr)
68	0·000	0·0
73	0·004	3·2
78	0·081	12·5
83	0·46	19·3
88	1·58	40·8
93	4·00	100·4
98	8·67	304·0
99	10·64	517·0
100 (saturated)	12·08	Infinity

saturation, is of great hydrologic significance. In this important case soil moisture is constant at saturation, so that equation (5) becomes extremely simple, namely

$$\frac{\partial Q}{\partial x} = i \qquad (6)$$

What this means in practice is that the near-saturated soil zones respond very rapidly to changes in rainfall intensity, even before overland flow begins. Most

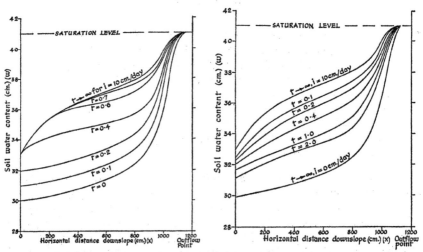

Fig. 5.1.4 Calculated moisture distributions in a 12-m-long soil trough, inclined at 22° during: (*left*) rainfall at a constant rate of 100 mm./day, and (*right*) drainage after indefinite rainfall of this intensity.

The data are calculated from the results of an experiment by Hewlett and Hibbert (1963).

important is the saturated zone which is often at the side of flowing streams, for the outflow from this zone is given by:

$$q_T = i \cdot x_s + q_T^* \tag{7}$$

where x_s is the width of the saturated zone;

$\quad q_T$ is the throughflow discharge per unit contour length;

$\quad q_T^*$ is the throughflow contribution from farther upslope; and

$\quad i$ is the rainfall after interception.

Equations (4) and (5) also provide a basis for assessing the influence of changing gradient or (with slight modification) of contour curvature on the throughflow discharge within the soil. The following four regions of a slope are the most likely to become saturated, and hence provide more rapid response to rainfall and more frequent saturation overland flow:

1. areas adjacent to perennial streams;
2. areas of concave upwards slope profile;
3. hollows (areas of concave outward contours);
4. areas with thin or impermeable soils.

3. Areas contributing to stream flow

The relative infrequency of overland flow in humid regions, together with the very low velocities of throughflow, suggest that most of the rainfall which falls on hillslopes is unable to reach a channel until long after the rainfall has stopped and the stream flood peak has passed. In other words, only water from a relatively small 'contributing area' is able to reach a channel in time to contribute to the flood hydrograph of the stream. In its simplest form, for a rainstorm of constant intensity, the contributing area, A_c, is defined as

$$A_c = \frac{\text{Stream discharge}}{\text{Rainfall intensity}} \tag{8}$$

Where the rainfall intensity varies during the storm its value in the equation is necessarily an average one, weighted towards the most recent intensities. The contributing area is continuously changing during the storm (fig. 5.1.5), and is generally at its greatest at about the same time as the peak discharge in the stream. For basins measured in North Carolina the maximum contributing area varied relatively little from storm to storm, but from basin to basin it ranged from 5 to 85% of the total drainage area. These measurements of contributing area can be compared with models derived from the two types of hillside flow: 1. 100% overland flow, and 2. 100% throughflow. During 100% overland flow water from all parts of the basin commonly reaches a channel within about an hour, so that for a storm lasting longer than an hour the contributing area, expressed as a percentage of the total drainage area, is

$$\frac{\text{Rainfall intensity} - (\text{Infiltration and surface losses})}{\text{Rainfall intensity}} \times 100\%$$

Under conditions where vegetation and soil are thin or absent, this contributing area may be large, and the value of 85% contributing area refers to an abandoned copper strip-mining area with less than 36% vegetation cover.

During 100% throughflow the contributing area consists of:

1. the area of the stream channels themselves, which is usually 1–5% of the drainage area;
2. the areas of saturated or near-saturated soil, mainly adjoining channels, which respond rapidly to changes in rainfall intensity as is shown in equations (6) and (7);
3. a narrow strip of hillside around the saturated areas, the width of which is determined by the slow rates of throughflow in unsaturated soils.

In the North Carolina basin with a 5% contributing area, the actual channel area and a swampy area backed up behind the stream-gauging installation

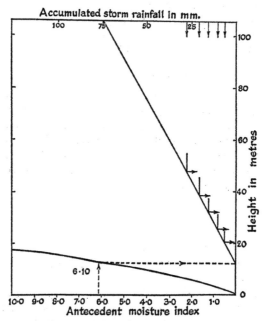

Fig. 5.1.5 An early concept of the variation of contributing area (considered as area below elevations shown), with initial moisture conditions and accumulated storm rainfall, for Bradshaw Creek, Tennessee (From T.V.A., 1964, Fig. 12).

accounted for almost the whole of the contributing area. Actual values of contributing area for humid drainage basins usually lie between these extremes, commonly in the 10–30% range, depending on soils, hillslope gradients, land use, and drainage texture.

Soils influence contributing area through their infiltration rate and the thickness of their permeable horizons, and attempts have been made to prove that contributing areas coincide with thin soil areas in a small drainage basin. Slope gradients influence the rate of throughflow (equation (4)), and hence the distribution of saturated soil and saturation overland flow. Vegetation cover and cultivation practices strongly affect the permeability of surface soil layers, mainly through the effect of vegetation in reducing rainsplash impact and through the effect of cultivation on depression storage and soil structure. In a basin with a high drainage density a relatively large area of hillslope is close to a channel, so that under throughflow the contributing area will also be relatively large; and under Horton overland flow there will be a relatively short time lag between the start of rainfall and a condition of maximum contributing area. Clearly these factors are not independent of one another, but each has a separate influence on contributing area.

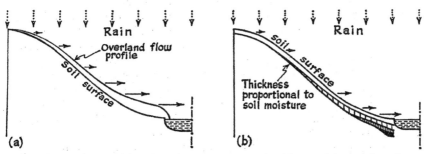

Fig. 5.1.6 Patterns of hillslope flow during Horton overland flow and throughflow. Arrow lengths show relative discharges over or through the soil.

(a). Horton overland flow (After Horton, 1945, p. 316). Thickness of water layer on surface is drawn proportional to actual thickness.

(b). Throughflow. Thickness of water layer below surface is drawn proportional to soil moisture content. Soil moisture from progressively earlier rainfalls is shown by progressively darker shading. The subsurface layer does not indicate the depth of infiltration into the soil.

Part of stream baseflow is derived from the water-table, which is itself supplied by deep percolation of water from the soil, but this contribution to baseflow is probably large only where well-defined aquifers are present. A large part of baseflow also comes from throughflow in the soil, which will take months to reach a channel from interfluve areas, and produces sufficient water to supply the measured baseflows in many areas. Since the same rain-water, much of it via throughflow, is responsible for both high and low flows in streams, all of the factors described above, which tend to produce high contributing areas during rainstorms, and hence a large proportion of total runoff during storms, also lead to reduced storage of rain-water after the flood flows have subsided, and hence to lower baseflows.

A final important contrast between the overland flow and throughflow models is that, whereas in overland flow it is the rain-water which is actually falling that flows into the stream during a rainstorm (fig. 5.1.6(a)), in throughflow, much of the water flowing into the channels is not physically the same as the rain-water which is currently falling (due to the time lag involved). It has been shown in infiltration experiments that almost all water flowing through the soil flows out in the order in which it flows in. This means that infiltrating water has to displace all of the soil water downslope before it can itself flow into the stream (fig. 5.1.6(b)), so that most water flowing into the stream, even at high flows, has been stored in the soil for a matter of weeks or months, and so has been able to come to chemical equilibrium with the soil. This soil water storage has obvious implications for interpreting the dissolved load of streams.

4. Summary

There are two extreme models of hillside water flow; the Horton overland flow model and the throughflow model. Horton overland flow occurs when rainfall intensity exceeds infiltration rate, and when it occurs at all in a basin, it is widespread. It is most common in semi-arid climates, and only occurs at progressively higher rainfall intensities under progressively thicker soil and vegetation covers. Throughflow occurs whenever the soil permeability decreases with increasing depth in the soil within some portion of the soil profile, most commonly at the base of the A horizon. Throughflow is probably the predominant mode of hillside flow in humid and humid–temperate areas, but it is of lesser importance under more arid or less-vegetated conditions with thin soils. When throughflow saturates the soil profile up to the surface, then saturation overland flow occurs. It occurs at much lower rainfall intensities than Horton overland flow, and is usually much more localized in its distribution, being commonest near streams. Under suitable conditions, both overland flow and throughflow may occur at any point although their relative frequencies will vary greatly from point to point. The separation of hillside flow into its two components, overland flow and throughflow, and a recognition of the distinct properties of each, allows a clearer understanding of the mechanisms of both streamflow and hillside erosion.

REFERENCES

BETSON, R. P. [1964], What is watershed runoff?; *Journal of Geophysical Research*, **69**, 1541–52.

BILHAM, E. G. [1936], Classification of heavy falls in short periods; *British Rainfall*, **75**, 262.

HEWLETT, J. D. and HIBBERT, A. R. [1963], Moisture and energy conditions within a sloping mass during drainage; *Journal of Geophysical Research*, **68**, 1081–7.

HORTON, R. E. [1945], Erosional development of streams and their drainage basins: hydrological approach to quantitative morphology; *Bulletin of the Geological Society of America*, **56**, 275–370.

HUDSON, N. W. and JACKSON, D. C. [1959], Results achieved in the measurement of erosion and runoff in Southern Rhodesia; *3rd Inter-African Soil Conference, Dalaba*, Paper No. 63.

JENS, S. W. and MCPHERSON, M. B. [1964], Hydrology of Urban Areas; In Chow, V. T., Editor, *Handbook of Applied Hydrology* (New York), Section 20, 45 p.

KIRKBY, M. J. and CHORLEY, R. J. [1967], Throughflow, overland flow and erosion; *Bulletin of the International Association of Scientific Hydrology*, **12**, 5–21.

LINSLEY, R. K., KOHLER, M. A., and PAULHUS, J. L. H. [1949], *Applied Hydrology* (New York), 689 p.

MUSGRAVE, G. W. and HOLTAN, H. N. [1964], Infiltration; In Chow, V. T., Editor, *Handbook of Applied Hydrology* (New York), Section 12, 30 p.

PHILIP, J. R. [1957–8], The theory of Infiltration; *Soil Science*, **83**, 345–57 and 435–48; **84**, 163–77, 257–64, and 329–39; **85**, 278–86 and 333–7.

TENNESSEE VALLEY AUTHORITY [1964], Bradshaw Creek – Elk River: A pilot study in area-stream factor correlation; *Office of Tributary Area Development, Knoxville, Tennessee*, Research Paper No. 4, 64 p. and 6 appendices.

TENNESSEE VALLEY AUTHORITY [1966], Cooperative Research Project in North Carolina: Annual Report for Water Year 1964–1965; *Division of Water Control Planning, Hydraulic Data Branch*, Project Authorisation No. 445.1, 31 p.

WHIPKEY, R. Z. [1965], Subsurface stormflow from forested slopes; *Bulletin of the International Association of Scientific Hydrology*, **10**, 74–85.

YARNELL, D. L. [1935], Rainfall intensity-frequency data; *U.S. Department of Agriculture, Misc. Publ. No. 204.*

5.II. Erosion by Water on Hillslopes

M. J. KIRKBY

Department of Geography, Bristol University

Water detaches and transports hillside material through the effect of raindrop impact; by unchannelled flow over the surface and within the soil; and by the formation and enlargement of a network of rills, gullies, and channels. Rainsplash detaches soil particles, and may on its own produce a net downhill transport of debris on a hillslope. Throughflow carries material in solution and in suspension within the soil mass, and in some cases carries material selectively along certain lines, which may be simply lines of greater permeability or may form small tunnels. Overland flow transports soil particles detached by rainsplash and may erode distinct channels, some of which may become a permanent part of the drainage channel network, whereas others are obliterated between storms by mass-wasting processes. Lateral and vertical cutting by pre-existing streams may erode and steepen the lower parts of slope profiles. These processes are the direct contribution of flowing water to hillslope erosion, but through the part played by soil moisture, water also enters indirectly into almost all other slope processes (Chapter 4.II). This section is concerned with a closer examination of the erosional processes caused directly by flowing water.

1. Surface and subsurface wash

On a level surface the splash-back following raindrop impact has been observed to move 4-mm stones distances up to 20 cm; 2-mm stones up to 40 cm; and smaller stones up to 150 cm. The stones are not transported in any one direction, but are part of a random exchange. On a slope the movement of individual stones is preferentially downhill, so that there is a net transport of material downslope. At any given rainfall intensity the *distance* of movement decreases with increasing particle size, but the *mass transport*, defined as the product of mean distance travelled and particle mass, *increases* with particle size for particles of up to 50 mm diameter (for rainfall intensities of 25–50 mm/hr). It is probable that there is not an indefinite increase, but a peak mass transport in the size range 100–400 mm diameter (which includes movement initiated by splash undermining). In areas with similar rainfall intensities the total rate of mass transport on a vegetation-free surface increases with mean annual rainfall. However, because vegetation cover also increases with mean annual rainfall, and vegetation shields the ground surface from rainsplash impact, mass transport does not increase indefinitely with mean annual rainfall but reaches a maximum and then decreases with further increase in rainfall (fig. 5.II.1). The maximum rate of 2·6 cm³/cm/yr occurs at a rainfall of 375 mm per year in the Western U.S.A. This rate may be compared with measurements of 0·09 cm³/cm/yr under a continuous vegetation cover in Britain, showing that rainsplash erosion is almost completely suppressed in humid areas except where there are local breaks in the vegetation cover.

Clays are certainly transported downwards through the soil from the A horizon to the B horizon by percolating water, and this is an essential part of soil development. As subsurface flow is predominantly lateral (throughflow), it is natural to assume that water also moves clays downslope, although there have

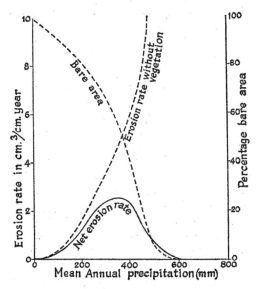

Fig. 5.II.1 Vegetation cover, erosion on bare ground, and erosion under natural vegetation, as each varies with mean annual rainfall in the south-western United States.

been few, if any, direct measurements of this process. In humid areas throughflow is more common than overland flow, but throughflow velocities are much less, so that it is not known which process is the more important. Dendritic networks of lines along which the permeability is greater and the A horizon is deeper are found, sometimes leading into the heads of stream channels. These lines carry greater throughflow and exhibit deeper weathering, and they may therefore be lines of more subsurface wash than in the surrounding soils, but these 'percolines' may also merely be the result of infilling of former extensions of the stream network. 'Piping', consisting of a subsurface network of small tunnels in soil and poorly cemented sediments, is widespread in semi-arid areas. Appreciable areas, especially on mesa tops, drain entirely through these piping systems, which supply springs lower down the hillside and carry all the material eroded from the areas they drain. Limestone cavern systems should perhaps be quoted as an extreme example of subsurface sediment transport.

2. Rill and gully erosion

As water flows over a hillside its depth varies in relation to the irregularities of the ground surface, being deepest in the depressions. It appears that most small depressions, even on an unvegetated hillslope, do not develop into linear channels, but any newly eroded channels that do form must begin with the con- centration of flow in some of these small depressions. Under conditions of wide- spread overland flow, a steady state will soon be set up in which overland flow discharge, on average, increases linearly downslope with increasing distance from the divide. Horton has shown how a critical distance from the divide may be calculated, at which the hydraulic power of the flow is great enough to over- come the strength of the soil and vegetation mat, and begin to erode a channel. There should therefore be a line of channel heads across a hillside, each channel head being at the critical distance from the divide. Falling from this line there should be a set of many sub-parallel rills, or shoe-string gullies, covering the hillside. This is Horton's classic runoff model, in which the critical distance from the divide is controlled by the overland flow intensity, the gradient, the hydrodynamic roughness of the surface, and the strength of the vegetation mat. A rill system of this kind does not generally last for very long, but is rapidly converted into a normal dendritic network through the dominance of a few rills which are enlarged to form gullies, each in its own small, eroded valley. Horton has described the process of cross-grading by which water overtops the small divides between adjacent rills and leads to progressive diversion of water into the rill which is at the lower level. This process can lead to the development of an overall gradient at an angle to the direction of flow of the rills, and the rills will gradually turn in direction so that they again flow down the line of steepest slope (Leopold, Wolman, and Miller [1964], Chapter 10).

Where rills exist cross-grading may be an important process of drainage inte- gration, but rill systems are not, in fact, common except on hillslopes with little or no vegetation cover. Schumm has shown that where they do exist they are often a seasonal phenomenon; the pattern he studied being destroyed each winter by frost action. Rill patterns may also be seasonally destroyed by field cultiva- tion, and both of these processes of rill destruction average the erosion of the rills across the width of the hillslope. Figure 5.II.2(a) shows how rill erosion, alternating with obliteration by tillage and surface wash, is tending to smooth out the irregular profile of a ploughed field into a smooth, convexo-concave form. Measurements show that the increasing total amount of debris transported with increasing distance downslope (fig. 5.II.2(b)) is achieved almost entirely by a rapid increase of transport in the gullies, and almost no increase between them. This is probably caused by a correspondingly rapid increase of discharge in the gullies and a progressive lateral diversion of overland flow into them from the interfluve areas.

Under rainfall conditions where Horton overland flow is being uniformly

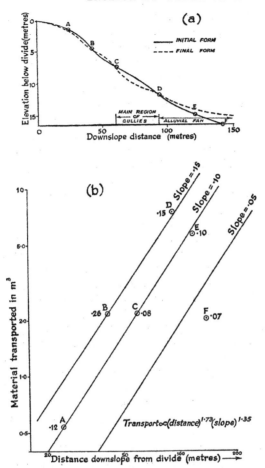

Fig. 5.II.2 Erosion and deposition by surface wash and small gullies in a ploughed field in Maryland, U.S.A., during the course of one year.

(a) Generalized longitudinal profile of field showing net erosion or deposition on same vertical scale.

(b) Correlation of transport rate (T) with distance from the divide (x) and tangent slope (s). Best-fit regression equation is $T \propto x^{1 \cdot 73} s^{1 \cdot 35}$, which is 'probably significant' at 95% level.

produced over a hillslope, the erosive force of the flowing water increases with both distance from the divide and with gradient. On a hillside of uniform gradient the erosive force is therefore greatest at the base of the slope; but on a convexo-concave slope the erosive force is greatest at a short distance downslope from the region of steepest gradient, so that it is here that Horton overland flow may be expected to initiate gullies. Saturation overland flow is preferentially produced in certain areas so that resultant gullying is also localized in these areas, namely; areas adjacent to perennial streams, areas of concave-upward slope profile, areas of thin or impermeable soils, and hollows (Chapter 5.1). These areas are strongly influenced by local factors of soil and three-dimensional valley shape, because the throughflow which allows saturation overland flow to develop travels so slowly that the water flows only a short distance while the rain is still falling. Distance from the divide is therefore not a control on the location of gullies which arise from saturation overland flow. Where Horton overland flow is involved, the factors are reversed in importance: with high flow velocities, distance from the divide is of paramount importance and, in comparison, local factors are much less important. A second distinction between the distribution of the two types of gullies is that Horton overland flow can initiate gullies which may be entirely separate from the main channel network, whereas saturation overland flow generally extends existing channel systems.

3. Slope profiles and drainage texture

The way in which overland flow and hydraulic erosion vary down a hillslope profile has important consequences for the form of the profile. Traditionally, convexities have been ascribed to soil creep and similar processes, and concavities have been ascribed to hydraulic processes, on the grounds that the erosive force of overland flow increases with increasing distance downslope. However, there is an effective increase of transporting capacity with distance only along water-eroded channels, and it is this that gives gullies their characteristic eroded, concave long profiles. On hillsides between channels the influence of distance from the divide is much less than has been formerly supposed for three reasons: 1. much of what has formerly been attributed to sheetwash is actually due to rainsplash, which is not influenced by distance from the divide; 2. even when overland flow is widespread, interfluve areas do not show a systematic increase of transporting capacity with distance downslope because a part of the flow is continually being diverted laterally into channels; and 3. where throughflow and saturation overland flow occur, distance from the divide is of little importance, since a steady state of flow is not achieved. For these three reasons, much of the force of the argument relating slope concavities to surface erosion by water is based on a premise which is, at best, partially true. Unchannelled hydraulic surface erosion is, therefore, like creep, likely to produce only convex hillslope forms.

Concavities at the bases of slopes must therefore be explained by mechanisms which do not rely on sheet erosion. There seem to be three possibilities: 1. Where a rill pattern is seasonally formed and destroyed, the movement of the rills from season to season allows their influence to be averaged across the slope so that the whole hillside is lowered uniformly at an average rate. Instead of the stream profiles becoming concave and the interfluves remaining convex, the changing positions of the rills lead to a general concavity. 2. A concentration of flow caused by convergence in a hollow will lead to rapidly increasing surface and subsurface flow, and a consequent progressive increase downslope in soil transport by all slope processes. The result will be a decrease in profile convexity, which may even become concave. 3. Any slope process which is in-filling a valley bottom will produce a concavity. In-filling may follow gullying, headward extension of existing streams, or lateral migration of streams. Slopes examined in the field show both depositional and erosional concavities.

The distribution of overland flow, through its effect on the exact positions of gully and stream heads, influences not only the form of the slope profile but also the texture of the whole drainage net. Since approximately half of the total length of streams in a basin is in first-order streams, a small change in the position of stream heads has a major effect on drainage density. Gullying appears to be the beginning of a vicious circle, since gullying produces a small valley which diverts more water into the gully, which in turn increases the rate of erosion. However, drainage densities do not increase indefinitely, because between major storms mass movement is able to partially refill the gullies, which are then able to carry more subsurface flow than before without saturation. Even under uniform conditions of agriculture and climate, the exact position of each stream head depends on the magnitude of the last major storm, and the period of time that has elapsed since. Thus, within a homogeneous area, neighbouring valleys may show different drainage densities reflecting a differing incidence of major storms, but indicating no difference in their basic equilibria. This variability of stream-head positions raises problems for the operational definition of stream networks, as the existence of eroded channels in the field is generally considered to be the basic criterion for comparison with maps and air-photography data.

4. Factors influencing rates of erosion

In determining the net erosional effect of hydraulic erosion, little work has been done on the relative contribution of each process. The bulk of the measurements have been of sediment collected either from artificially sprinkled plots in cultivated fields, or from natural streams. Table 5.II.1 summarizes some representative data from soil-erosion experiments, and shows that average rates of soil stripping may be as high as 30 mm/yr. The values are perhaps less important than the factors which control the rate of soil erosion. These factors fall within four main groups: 1. the depth, permeability, and other properties of the soil; 2. the land gradients; 3. the frequency distribution of rainstorms which are able to produce overland flow; and 4. land use, especially the amount of vegetation cover during the rainstorm periods of the year. In these plot experiments land use and soil properties are to some extent within the control of the experimenter, whereas stream measurements of erosion are usually done for larger areas and under more natural conditions, in which vegetation and soil are strongly correlated with climate. The measured rates of stream erosion are therefore expressed in terms of climatic and topographic variables. Langbein and Schumm [1958] have correlated erosion in basins of 50–5,000 km² with rainfall, across the United States, and have shown that there is a maximum erosion at a rainfall of about 250–350 mm p.a. (fig. 5.II.3). Evidence from both humid and arid regions suggests that most stream sediment is derived directly from its banks (as much by mass movement as by hydraulic action), and this is in accordance with the

TABLE 5.11.1 Measured rates of soil erosion from experimental plots

Area	Mean annual rainfall (mm)	Vegetation cover	% runoff	Soil loss (mm/yr)	Source
South-east U.S.A.	2,500–4,000	Oak forest	0·8	0·008	Meginnis [1935]
		Bermuda grass pasture	3·8	0·030	
		Scrub-oak woodland	7·9	0·10	
		Barren abandoned land	48·7	24·4	
		Cultivated: rows on contour	47·0	10·6	
		Cultivated: rows downslope	58·2	29·8	
Rhodesia	1,000	Dense grass	2·7	0·018	Hudson and Jackson [1959]
		Bare ground	38·0	2·3	

Erosion $\propto$ (Slope)^a × (Distance)^b × (Rainfall energy factor) × (Soil erodibility factor)

Slope exponent, $a = 0·6 - 2·0$

Distance exponent $\begin{cases} b = 0·0 - 0·7 \text{ without gullying} \\ b = 1·0 - 2·0 \text{ with gullying} \end{cases}$

decrease of drainage density with increased vegetation cover and infiltration rate which has been observed in the United States in regions of more than 250–350 mm of rainfall p.a. (fig. 5.11.4).

It seems that the principal cause of high erosion rates in semi-arid regions is the relative lack of vegetation cover, which leads to low infiltration rates. The result is increased overland flow, which in turn leads to high drainage densities and to an increase of runoff and erosion. Within humid areas the completeness of the vegetation cover varies relatively little with areal differences in rainfall. Therefore, as one moves from dryer to wetter areas (starting from 250–350 mm p.a. for the United States) the rates of erosion initially decrease rapidly to the point (about 600 mm p.a. for the United States) at which a total vegetation cover is established, and thereafter change very little (fig. 5.11.3). Despite the close correlation with annual rainfall, it is the distribution of rainfall within the year, much more than the annual total, which controls the vegetation and amount of erosion. More complex causal chains have been proposed, in which low-intensity rainfalls are said to have most influence on vegetation cover, whereas high-intensity rainfalls have most influence on peak stream runoffs. Thus a climatic change from low-intensity winter rains to high-intensity summer rains (as has occurred in the American South-west) will significantly increase rates of erosion, even though the change in mean annual rainfall is slight.

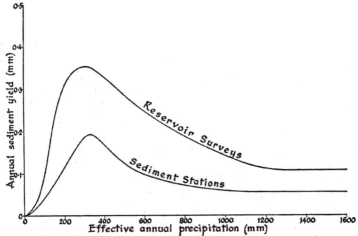

Fig. 5.11.3 The variation of total basin erosion rate with effective mean annual precipitation in the United States, for river sediment stations (of average area of 4,000 km²) and reservoir surveys (of 80 km² average area). The difference illustrates the normal variation of sediment yield ∝ (basin area) $^{-0.15}$ (After Langbein and Schumm, 1958).

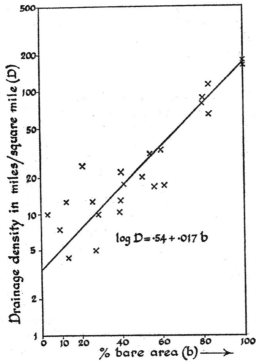

Fig. 5.11.4 The variation of drainage density measured in the field with the percentage of bare, unvegetated area, for third- and fourth-order basins in the United States (After Melton, 1957).

Factors which influence the areal pattern of erosion through climate and vegetation also influence the temporal pattern of erosion as conditions change at a point. There is evidence of alternate periods of gullying and valley deposition in many different climatic regions. These alternations are most striking in semi-arid regions like the American South-west, but they are also present in temperate regions like the eastern United States and Britain. Work in semi-arid areas shows fairly good correlation between dry periods and periods of valley cutting, although the most recent period of valley cutting in the south-western United States, around 1900, has been ascribed to either a minor climatic change or, alternatively, to destruction of vegetation through over-grazing.

In more humid regions fluctuations in climate are not usually of sufficient magnitude to destroy vegetation through aridity. The role of man, through clearing the vegetation for cultivation, has therefore been relatively greater. Experiments have shown that the greater the rainfall and the thicker the natural vegetation cover (up to a point), the greater will be the acceleration of erosion consequent on stripping the vegetation. However, this effect is offset by the more rapid regeneration of vegetation in humid climates, and by the generally greater thickness of soil which must be stripped before unweathered parent material is exposed. Perhaps for these reasons the soils most sensitive to erosion appear to be in areas which are semi-arid, and in these areas it is most common for irreversible soil erosion to occur, leading to bedrock slopes on which vegetation cannot effectively regenerate.

5. Summary

To summarize the current state of knowledge of hillslope fluvial erosion processes; it appears that subsurface wash is of unknown, but probably minimal, importance; surface wash between channels is produced mainly by raindrop impact, is very dependent for its efficiency on local or general absence of vegetation, and is also of minor importance. Only the erosive power of water flowing in clearly-defined channels is a truly effective transporting and eroding agent. The positions of channel heads, which control the drainage texture, vary areally between basins and fluctuate through time within a basin. At any moment their positions depend on the incidence of major storms which are extremely local in their effect; but the average positions of channel heads over a period of time are controlled mainly by soil, slope gradient, precipitation, and vegetation cover. Man can interrupt the natural balance by his control of vegetation and, to a lesser extent, by tillage, and most agriculture would lead to gullying if no preventive measures were taken. However, the extent to which conservation measures are necessary varies widely, according to the extent to which overland flow is increased by cultivation. Vegetation is seen to play a major role in the rate at which hillslopes are eroded by running water, and the influence of both man and climatic change is chiefly through their effect on changing the vegetation cover.

REFERENCES

BRYAN, K. [1928], Historic evidence of changes in the channel of the Rio Puerco, a tributary of the Rio Grande in New Mexico; *Journal of Geology*, **36**, 265–82.

BUNTING, B. T. [1964], Slope development and soil formation on some British sandstones; *Geographical Journal*, **130**, 73–9.

CARLSTON, C. W. [1966], The effect of climate on drainage density and streamflow; *Bulletin of the International Association of Scientific Hydrology*, **11**, 62–9.

ELLISON, W. D. [1945], Some effects of raindrops and surface flow on soil erosion and infiltration; *Transactions of the American Geophysical Union*, **26**, 415–29.

FLANNERY, K. V., KIRKBY, A. V. T., KIRKBY, M. J., and WILLIAMS, A. W. [1967], Farming systems and political growth in Ancient Oaxaca; *Science*, **158**, No. 3800, 445–54.

HACK, J. T. [1942], The changing physical environment of the Hopi Indians of Arizona; *Reports of the Awatovi Expedition, Peabody Museum, Harvard University*, Report No. 1, 85 p.

HACK, J. T. and GOODLETT, J. G. [1960], Geomorphology and forest ecology of a mountain region in the central Appalachians; *U.S. Geological Survey Professional Paper 347*, 66 p.

HORTON, R. E. [1945], Erosional development of streams and their drainage basins: hydrophysical approach to quantitative morphology; *Bulletin of the Geological Society of America*, **56**, 275–370.

HUDSON, N. W. and JACKSON, D. C. [1959], Results achieved in the measurement of erosion and runoff in Southern Rhodesia; *3rd Inter-African Soil Conference, Dalaba, Paper No. 63*.

LANGBEIN, W. B. and SCHUMM, S. A. [1958], Yield of sediment in relation to mean annual precipitation; *Transactions of the American Geophysical Union*, **39**, 1076–84.

LEOPOLD, L. B. [1951], Rainfall frequency, an aspect of climatic variation; *Transactions of the American Geophysical Union*, **32**, 347–57.

LEOPOLD, L. B., WOLMAN, M. G., and MILLER, J. P. [1964], *Fluvial Processes in Geomorphology;* (Freeman & Co., San Francisco), 522 p.

MELTON, M. A. [1957], An analysis of relations between climate and landforms; *Office of Naval Research, Technical Report 11; Project N.R. 389–042*, 102 p.

MELTON, M. A. [1958], Correlation structure of morphometric properties of drainage systems and their controlling agents; *Journal of Geology*, **66**, 442–60.

PARKER, G. G. [1963], Piping, a gemorphic agent in landform development in the drylands; *International Association of Scientific Hydrology*, Publication No. 65, 103–13.

SCHUMM, S. A. [1956], Evolution of drainage systems and slopes in badlands at Perth Amboy, New Jersey; *Bulletin of the Geological Society of America*, **67**, 597–646.

SCHUMM, S. A. [1964], Seasonal variation of erosion rate and processes on hillslopes in western Colorado; *Zeitschrift fur Geomorphologie*, Supplementband **5**, 215–38.

6.II. The Geomorphic Effects of Ground Water

PAUL W. WILLIAMS

Department of Geography, Trinity College, Dublin

Ground water will be considered here as all water of meteoric origin within the soil and underlying rocks. Thus defined its distribution is widespread, but since its geomorphic effects operate primarily through solution, its influence is most striking in carbonate terrains. This brief account of the role of ground water in modifying the relief will therefore focus attention on limestone and dolomite landscapes where solution manifests itself through *karst* landforms.

The karstic process of solution and concomitant rock settling and collapse is largely directed by geological constraints acting on the passage of subsurface water. A brief examination of the principles of ground-water movement through limestones is therefore essential for proper understanding of the resultant topography.

1. Ground water in limestones

The characteristic dryness of limestone areas is a consequence of the highly permeable nature of the rock. This permeability is dependent upon the volume and interconnection of both primary and secondary interstices in the rock, and while the most permeable limestones possess both primary and secondary porosity, the latter is usually more significant for fluid circulation because of the larger voids concerned.

The susceptibility of limestone to solution is fundamental in the enhancement of its permeability, for the stimulation of solution provided by the very act of water circulation is instrumental in the progressive enlargement and modification of the subterranean conduit system. The capacity of carbonates for this progressive development of permeability distinguishes them hydrologically from most other rocks, and in large measure accounts for the distinctive quality of their topography – above and below ground. However, it is clear that water conditions in limestones vary considerably from place to place as lithology, structure, and geomorphic history change. So it is impossible to formulate a neat, all-embracing theory of karst hydrology that may be used to assist in the explanation of surface landforms. Current opinions on the nature of water conditions in limestones waver basically between the ideas of Martel [1894] and Grund [1903], namely between discrete conduit and interlinked network approaches. Anglo-American hydrologists have always favoured the latter, since King's [1899] theory of ground-water motion and Davis's [1930], Swinnerton's

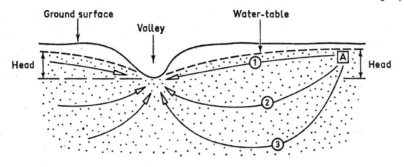

Water at A will divide itself into proportions directly related
to the ease of movement. In the three directions illustrated,
the largest amount will take the shortest and least resistant
route, path 1, and the smallest amount path 3.

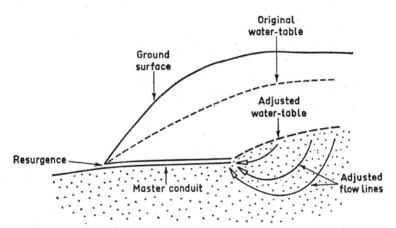

Fig. 6.II.1 *Above:* Water movement beneath the water-table in limestones, after
Swinnerton (1932).
Below: Water conduit development and water-table adjustment in limestones, after
Rhoades and Sinacori (1941).

[1932], and Rhoades' and Sinacori's [1941] adaptation of it for limestone terrains
have emphasized the concept of a unified body of ground water, with the upper
limit of the permanently saturated zone being delimited by a hypothetical surface, the water-table (fig. 6.II.1). However, in continental Europe conditions
leading to the existence of simple water-table situations have never been fully
admitted, for study of the hydrology of the limestone Alps has constantly revealed intricate ground-water systems in which the water-table concept has no
apparent viability (see Zötl [1965] for a summary of some valuable modern work).

Although hypotheses on the nature of limestone hydrology and corresponding cave genesis are almost as diverse as the terrains which stimulated them (reviewed in Warwick [1962]), there is still much agreement on the general nature of water movement in the aerated or vadose zone, and this is what concerns us most, for it is there that solution has its greatest effect on surface landforms. In the vadose zone water can be divided into that which percolates through the soil and fissures in the underlying rock, and that which passes rapidly underground via swallow holes and caves. The two categories will be referred to as percolation water and vadose stream water respectively. Drainage is basically vertical towards the zone of phreatic water, the third category, where the rock is permanently saturated and where lateral flow becomes more important.

2. Topographic effects of ground-water solution

Underground water is able to affect topography because of its solvent capacity. The majority of investigators agree that the most important solvent in carbonate terrains is carbonic acid, which is produced by the solution of carbon dioxide from the air, and particularly from the soil atmosphere, where the partial pressure of the gas is usually greater, although numerous other natural solvents are known to exist (Keller, 1957).

Ground water in limestones was subdivided above into percolation water, vadose stream water, and phreatic water. Each has a different erosive role and a distinctive geomorphic effect (fig. 6.11.2). Percolation water infiltrates very slowly into the limestone mass, and hence its corrosive capacity or 'aggressivity' is often spent within a few metres of the surface (Gams, 1966; Pitty, 1968; Williams, 1968). Subterranean stream water, on the other hand, moves rapidly underground, and its solvent capacity may not be satiated until a depth of several hundred metres or more is reached, and in addition to their corrosive role, vadose streams also erode mechanically like their surface counterparts. The activity of phreatic water is difficult to ascertain because of inaccessibility. Swinnerton [1932] and Rhoades and Sinacori [1941] (fig. 6.11.1) postulated maximum solution and cave formation at the level of the water-table where ground-water flow is concentrated in its greatest volume. Their ideas were strongly supported by Kaye [1957], who experimentally showed solution to be proportional to the velocity of solvent motion. Theoretical support for deep subterranean solution has been provided by Bögli [1964], who has indicated how mixing waters with different carbonate concentrations are imparted renewed capacity to dissolve even if the original solutions were saturated.

The morphological effects of all styles of ground-water solution tend in the same direction, namely ultimate overall lowering of the relief, and numerous estimates of this have been made in different countries, the majority falling in the range 0·02–0·1 mm/yr. Nevertheless, the short-term effects of the above three categories of ground water differ. Percolation solution directly lowers the surface because it operates most strongly in the uppermost few metres, but its intensity varies according to geological factors, such as fracturing, which concentrates percolation routes. Zones of high-frequency jointing are therefore especially

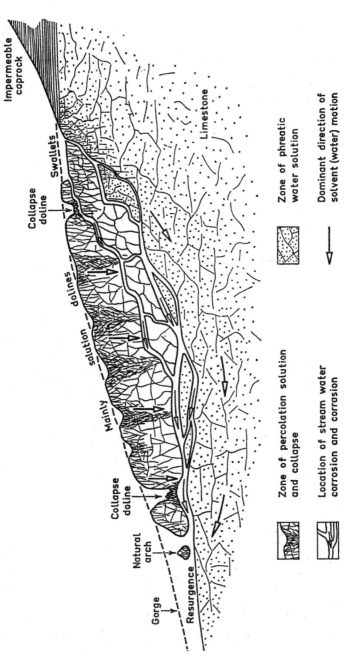

Fig. 6.II.2 The distribution of solution through a limestone mass.

Impermeable caprock

Swallets

Collapse doline

dolines

solution

Mainly

Collapse doline

Natural arch

Gorge

Resurgence

Limestone

Zone of phreatic water solution

Dominant direction of solvent (water) motion

Zone of percolation solution and collapse

Location of stream water corrosion and corrasion

prone to attack, and in such locations solution depressions develop (fig. 6.11.3). However, the considerable variation in the density and form of these features in tropical, temperate, and sub-arctic zones suggests that the overriding consideration in their distribution and morphology is climatic. Percolating water passing through thick drift (especially glacial and alluvial) overlying limestone is also frequently responsible for myriads of small enclosed depressions entirely within the deposit: *drift dolines*[1] (fig. 6.11.3). The percolating water erodes the drift partly by selective solution of its contents and partly by downwashing of fines

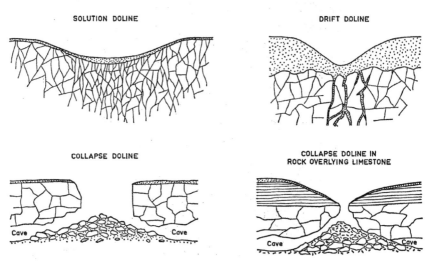

Fig. 6.11.3 Types of doline.

into voids in the underlying rock. This important but frequently overlooked process is known as chemical and mechanical *suffosion*.

It is not yet possible to compare quantitatively the erosive effects of percolation and vadose stream waters, but qualitative differences are immediately apparent. The most obvious contrast is in the distribution of denudation, which in the first case is relatively uniform over the limestone outcrop, but in the second is restricted to the trace of the stream. Vadose stream action, be it corrosion, corrasion, or the pressure effects of flood-water, results in the undermining of limestone outcrops and in the production of steep-sided collapse-induced depressions on the surface that contrast markedly with the rounded forms of percolation water solution depressions (fig. 6.11.3). Collapse depressions frequently have elongate plans (as do solution depressions if there is a particularly dominant joint direction) which may cut across the structure where subterranean streams are followed. But depressions are only produced by collapse above a stream where the cave roof is thin enough to permit the transmission to the surface of roof falls. The critical roof thickness is a function of the span of the

[1] The author now prefers to use the term *subsidence dolines*.

passage, the magnitude of discharge fluctuations in the stream, the mechanical competence of the overlying beds, and the degree to which they are corrosionally weakened. As a rough guide, a cave roof thicker than about 100 m is usually adequate to prevent sporadic collapse effects reaching the surface.

Although phreatic solution is difficult to measure, it has been shown to be important in the early stages of development of many caves (Bretz, 1942). Yet while the morphological effects of phreatic solution are widespread underground, there are few surface manifestations. The main geomorphic effect of deep ground-water solution is to reduce the overall volume of the limestone mass.

3. Ground-water solution and river basins

The progressive enlargement of voids in limestone by ground water results in automatic lowering of the hydrostatic equilibrium level. As this continues, surface streams lose more and more water into their beds until eventually they pass permanently underground. The morphological functions of the lower courses of such beheaded rivers inevitably change, particularly since the engulfed headwaters may not later rejoin their original basins. Subterranean capture in limestone areas is common where rapid water-table lowering is being induced by the entrenching of major through streams; the most favourable position for piracy being where two neighbouring rivers are incised to different levels. In such cases water is diverted underground through the topographic divide to the lower system (fig. 6.11.4). Self-piracy, leaving dry ox-bows, may also occur in meandering streams, for the hydraulic gradient through a meander neck is steeper than down the longer stream course (fig. 6.11.4); so subterranean meander cut-off is encouraged.

4. Surface features associated with subterranean streams

A variety of landforms develop from a combination of surface- and ground-water processes at the points where streams pass underground. The simplest and most widespread feature is the *swallow hole* (stream-sink); this is a small enclosed basin into which a stream flows and passes underground. At the site of ingress the water of a sinking stream is normally aggressive; thus morphological evolution is rapid. Corrosion acts directly and indirectly by widening fissures and inducing collapse. As swallets enlarge, they develop into elongate closed basins with steep downstream ends, more properly known as *blind valleys*. Processes operating in them are similar to those in swallets, and Gams suggested that in northern Yugoslavia their length and width is inversely related to the hardness of inflowing streams. Considerable further enlargement under conditions favourable to flooding, and hence lateral solution plantation, will develop a blind valley into a *karst margin polje* (marginal polje), although such landforms are attributable more to surface corrosion than ground-water action.

The cave roof downstream of a swallow hole is usually relatively thin, and since the zone is subject to intense chemical weathering and violent pressure changes with fluctuating stream discharge, collapse depressions frequently occur even where the cave passes beneath overlying, non-limestone formations (fig.

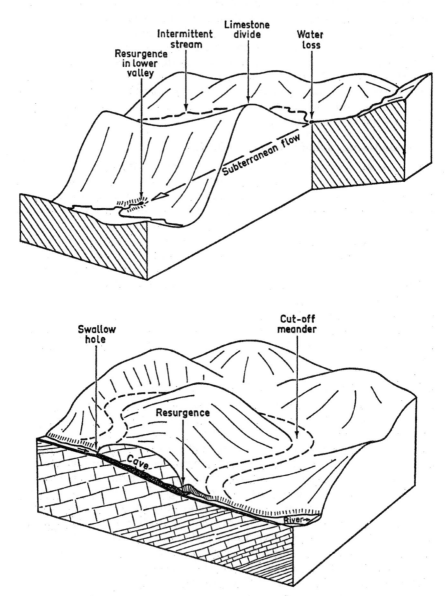

Fig. 6.11.4 Examples of (*above*) subterranean river capture, and (*below*) subterranean meander cut-off.

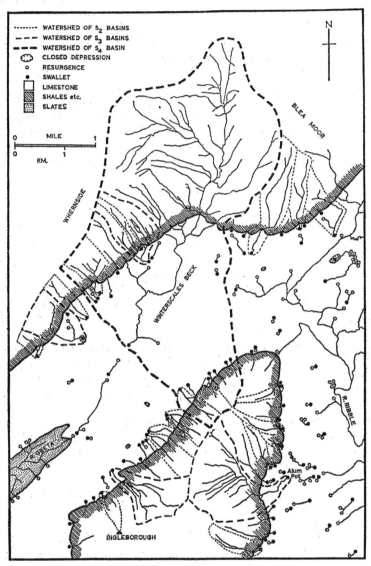

Fig. 6.11.5 Some karst features of the Ingleborough district, Yorkshire, England (From Williams, 1966).

6.11.3). As the stream passes farther and more deeply underground, associated collapse forms on the surface are less frequent, although in shallow systems, the river may be traced for considerable distances. However, subterranean streams do not always maintain discrete courses, especially where they flow into a phreatic network; so surface collapses cannot always be related to simple cavern plans.

At resurgences the morphological situation is in some respects similar to that at swallow holes, but in reverse. An increasing frequency of collapse depressions often heralds the emergence of an underground stream. The point of resurgence may be a *steep-head*, marked by steep or even overhanging slopes which retreat vigorously upstream by undermining and collapse. Downstream of the steep-head, the river commonly flows in a gorge, and cave-roof remnants may form natural arches. Gorges are frequently found where ground water resurges, although in karst areas they are not always produced by cavern collapse.

5. The quantitative analysis of karst landforms

The morphometric approach to the accurate description of individual karst landforms and of their interrelationships (Williams, 1966; LaValle, 1967), although in its infancy, is already casting doubt on the traditional notion that karst landscapes are chaotic and disorganized. Morphometry, as developed from map and air-photograph analysis, is able to describe only the more gross properties of landforms, whereas the more subtle attributes can be observed and measured only in the field. However, even there landform identification is by no means always simple, and the majority of dolines, for example, cannot safely be assigned to a solution or collapse category without examination by excavation. The special nature of karst, where solution and collapse combine in varying proportions to produce a continuum of landforms, renders accurate genetic identification extremely difficult. The morphometric approach must therefore abandon the majority of field-worker's techniques and concentrate on measurable properties of shape, size, and distribution.

In any estimation of the geomorphic effects and morphological role of ground water it is thus advisable first to try to gain some idea of the general nature of the ground-water system itself, as follows (it being assumed that the limestone mass is pure throughout):

1. From a detailed geologic map of the district under examination, delimit the limestone outcrop, measure its area (A_l) (12 square miles)[1] (fig. 6.11.5), and give its thickness (H_l) (750 ft).
2. Locate swallow holes (S) and order them (S_o) (fig. 6.11.6) according to the Strahler order of the stream flowing into them ($\Sigma S_1 = 31$; $\Sigma S_2 = 19$; $\Sigma S_3 = 4$; $\Sigma S_4 = 1$; plus 37 which cannot be ordered). Note their altitudes (H_{so}) ($\bar{H}_s = 1,127$ ft) and distance to their neighbours having the same order (L_{so}). ($L_{s1} = 0.22$ miles; $L_{s2} = 0.43$ miles; $L_{s3} = 1.66$ miles).

[1] The values given for the morphometric parameters relate to 1 : 25,000 maps of the Ingleborough district shown in fig. 6.11.5.

3. Locate every karst spring (rising: K_r) ($\Sigma K_r = 81$) within the limestone outcrop, its altitude (H_r) ($\bar{H}_r = 1,009$ ft) and distance to its nearest neighbour (L_r).
4. Mark the course and calculate the length (L_i) of every through-flowing stream.

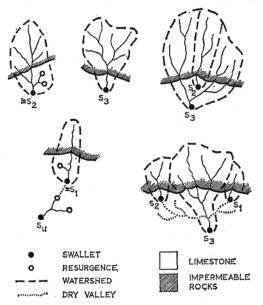

Fig. 6.11.6 Procedure for ordering Swallets and delimiting their catchments (From Williams, 1966).

Given this information, the following attributes of the karst drainage system can be estimated – bearing in mind that the larger the area, the more generalized the information derived:

1. Swallet density $\left(D_s = \dfrac{\Sigma S}{A_l}\right)$ (7·67 per square mile); a measure of the number of streams draining into the limestone per unit area.

2. Rising density $\left(D_r = \dfrac{\Sigma K_r}{A_l}\right)$ (6·57 per square mile); a statement of the number of risings per unit area.

3. The swallet/rising ratio $\left(R_{sr} = \dfrac{\Sigma S}{\Sigma K_r}\right)$ (1·12); indicating the amount and direction of stream branching underground. Purely vadose systems are expected to have R_{sr} values of >1, while areas with important phreatic ground-water bodies will have lower values which could descend to <1.

4. Mean shortest distance of underground flow (L_u) (0·36 miles); the mean straight-line distance between each swallet and its nearest downhill resurgence or river channel.

5. The *vadose index* ($V_i = \bar{H}_s - \bar{H}_r$) (118 ft); a measure of the depth of the zone of aeration (fig. 6.11.7). This is significant only for small areas, for the hydrostatic equilibrium level (which should determine $\bar{H}_r$) varies according to subterranean evolution, structure, and lithology.

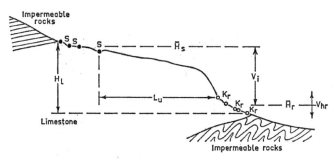

Fig. 6.11.7 The relationships of some karst parameters.

6. The *rising coefficient* $\left(V_{hr} = \dfrac{\sigma H_r}{\bar{H}_r} \times 100\right)$ (12·5); where σ is the standard deviation estimated from the sample. This defines the variation in altitude of springs; so in areas of pure limestone it is a direct measure of the uniformity of the water-table surface. It is thus an indirect measure of the state of evolution of the ground-water network.

7. *Stream density on limestone* $\left(D_l = \dfrac{\Sigma L_l}{A_l}\right)$ (2·4 miles per square mile); is a statement of the length of permanent streams per unit area, and is thus an indirect measure of permeability in an area of uniform climate.

8. Semi-logarithmic plots of order against mean number of swallets, mean distance to swallets of the same order ($\bar{L}_{so}$), and mean area of swallet basins ($\bar{A}_{so}$) summarize other attributes of swallet systems (fig. 6.11.8).

Where additional information is available from hydrologic surveys, bore-hole data, etc., it may be possible to locate phreatic water divides and so delimit ground-water drainage systems. Further characteristics of these systems may be estimated using an index of *relative karst system relief* (H_k) (this for any locality is the difference in height between the basin's lowest resurgence point and the highest place in the area considered) and a *karst relief ratio* (R_k) (this for a given site is the ratio of karst system relief to the distance of the site from the nearest phreatic divide) (LaValle, 1967). The relative karst system relief is related to the vadose index, since it measures the maximum depth of the zone of aeration in any particular area. The karst relief ratio was designed to estimate hydraulic gradient, but its effectiveness is questionable.

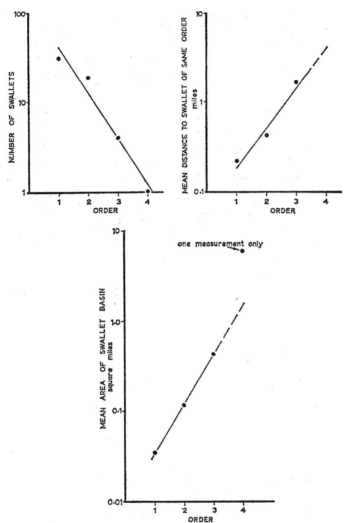

Fig. 6.11.8 Swallet relationships in the Ingleborough districts, Yorkshire, England (From Williams, 1966).

6. Closed depressions

Further analysis should also be made of surface conditions, which in a karst area implies particular attention to closed depressions (swallow holes, solution dolines, collapse dolines, etc.), including any enclosed hollow from the smallest pit to the largest karst polje. But since ground-water solution plays a subservient role to lateral surface-water solution in the origin of poljes, our concern here is only for the smaller features.

A temperate-zone doline is usually a relatively simple basin set in a gently rolling limestone surface, and, even in alpine areas, it is delimited on a map by a break of slope at its rim, indicated by a scar symbol or closed contour. However, many tropical depressions, with the exception of swallow holes, are different in form and setting. Whereas temperate depressions are often circular or ovoid in plan, the contours of their tropical counterparts are frequently star-shaped. The tropical basins are also interspersed between steep-sided, residual hills from which storm runoff drains into the intervening hollows. The hillsides are thus integral parts of the basin. While temperate depressions are adequately defined by their bounding break of slope, tropical depressions of the cockpit type seem best defined by their topographic divides (fig. 6.11.9); both methods in fact demarcate the immediate catchment areas of the closed basins concerned.

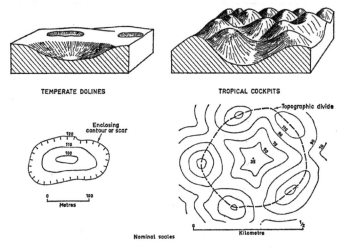

Fig. 6.11.9 Delimitation of temperate and tropical closed depressions.

Having established the immediate catchment area as the most realistic criterion for delimitation of karst depressions, the following analytical procedure may be followed:

1. Locate and count all closed depressions (C_d) (95); measure their areas (A_{cd}); maximum relief (difference between lowest point in basin and highest point on rim) (H_{cd}); length/width ratios (R_{de}); and long axis orientations in degrees of azimuth relative to the direction of general land slope (θ_s), also where appropriate, with respect to the geological strike direction (θ_l), major joint trend (θ_j), and maximum hydraulic gradient (θ_h).

2. Divide the area under examination into convenient sized units (e.g. 1 km²) and calculate for each unit

   (a) Closed depression density $\left(D_{cd} = \dfrac{\Sigma Cd}{A_l}\right)$ (7·8 per square mile); a simple statement of depressions per unit area.

(b) Mean area of depressions $\left(\bar{A}_{cd} = \dfrac{\Sigma A_{cd}}{N}\right)$ (approximately 0·0005 square miles).

(c) Mean depression relief $\left(\bar{H}_{cd} = \dfrac{\Sigma H_{cd}}{N}\right)$

(d) Index of pitting $\left(R_p = \dfrac{A_l}{\Sigma A_{cd}}\right)$ (2,500); indicating the extent to which the surface is dissected with depressions – unity representing complete pitting.

(e) Mean elongation ratio of depressions $\left(\bar{R}_{de} = \dfrac{\Sigma R_{de}}{N}\right)$.

(f) Mean depression orientation relative to generalized land slope, geological features, and hydraulic gradient.

From the above calculations, spatial variations will become apparent which require explanation with particular attention to ground-water influence. At this stage it should be stressed that here we pass largely into the realm of hypothesis, since as yet there has been little quantitative substantiation of the possible relations that are suggested.

Closed depression density over an area with relatively uniform climate, limestone lithology, structure, and slope should show little variation, because percolation solution should be fairly even. But should above-average depression densities occur, they may be interpretable as due to a greater incidence of collapse, especially if the higher densities are found round the borders of the limestone outcrop near swallets and resurgences. The possible effect of ground-water-induced collapse may therefore be tested by exploring the relation between closed depression density per unit area and mean depth to water-table per corresponding unit area. The level of the water-table (where it exists) should be significant, as it is there that cave passage formation, with concomitant roof collapse, is sometimes thought to be most rapid. A negative correlation should be expected where collapse-induced depressions are important. If depth to the water-table is unknown it can be approximated as varying with distance from the phreatic divide, resurgences, or limestone edge. Where lithology is not uniform, the influence of other variables, such as rock purity and jointing, may be more important determinants of depression density. In order to assess the relative significance of a range of variables on closed depression density, a multiple correlation programme will be required.

Closed depression shape and orientation may also reflect phreatic and vadose stream influence, but such control may only stand out if elongate depressions transgress the structure. Those accordant with structure could result from any combination of solution and collapse, and only field examination is likely to determine their evolution. The proportions of aligned and non-aligned depressions with respect to various aspects of structure should therefore give a minimal estimate of the importance of collapse, but more accurate would be depression orientation in relation to the direction of maximum hydraulic gradient.

Judging from LaValle's work in Kentucky, phreatic water control seems also reflected in the inverse relations of both depression elongation and percentage of structurally aligned depressions with distance from resurgence mouths. An explanation of this apparent causal relation might be that the magnitude of cave passages and corresponding roof falls decrease upstream while roof thickness increases (until, of course, the swallow holes are reached); thus the chances of collapse effects modifying surface landforms diminish upstream until the inflow zone is approached. It has also been suggested elsewhere that the hydrostatic equilibrium level has a direct effect on tropical karst relief, but this relation, like so many others in karst morphometry, is statistically unproven.

7. Conclusions

Karst landscapes are largely the result of solution on the surface and underground, plus collapse induced by solution. The dimensional characteristics and spatial distribution of karst landforms are being quantitatively described and a sensible if complex arrangement of features is being revealed; not the utter chaos that was once ascribed to karst. The techniques involved in this analysis are in the early stages of development, and many possible interrelationships remain to be tested. The problems of isolating and quantifying the landform responses to percolation solution, vadose stream action, and phreatic water solution are very great, and indeed attempts to do so are rather premature at this early stage in karst morphometry, when general patterns and the full range of related landforms have yet to be adequately described.

The writer chose to illustrate the geomorphic effects of ground water by concentrating on karst, not only because of the limits of personal experience but also because in the shaping of karst landscapes ground-water processes are more dominant than any other. Nevertheless, the importance of ground water in the evolution of other landform systems must not be under-estimated, particularly in the tropics, where ground-water weathering (Berry and Ruxton, 1959; Ollier, 1960; Thomas, 1966) and ground-water deposition (Woolnough, 1927; Prescott and Pendleton, 1952; Wopfner and Twidale, 1967) are fundamental processes in the evolution of the landscapes.

Acknowledgement. The author is grateful to Mr J. N. Jennings of the Australian National University for valuable criticism of the manuscript.

REFERENCES

BERRY, L. and RUXTON, B. P. [1959], Notes on weathering zones and soils on granite rocks in two tropical regions; *Journal of Soil Science*, **10**, 54–63.
BÖGLI, A. [1964], Mischungskorrosion – ein Beitrag zum Verkarstungsproblem; *Erdkunde*, **18**, 83–92.

BRETZ, J. H. [1942], Vadose and phreatic features of limestone caverns; *Journal of Geology*, **50**, 675–811.

DAVIS, W. M. [1930], Origin of limestone caverns; *Bulletin of the Geological Society of America*, **41**, 475–628.

GAMS, I. [1962], Blind valleys in Solvenia (summary in English); *Geografski Zbornic*, **7**, 265–306.

GAMS, I. [1966], Factors and dynamics of corrosion of the carbonatic rocks in the Dinaric and alpine Karst of Slovenia (Yugoslavia); *Geografski Vestnik*, **38**, 11–68.

GRUND, A. [1903], Die Karsthydrographie: Studien aus Westbosnien; *Geographische Abhandlungen herausgegeben von A. Penck*, **7**, 103–200.

KAYE, C. A. [1957], The effect of solvent motion on limestone solution; *Journal of Geology*, **65**, 35–46.

KELLER, W. D. [1957], *The Principles of Chemical Weathering* (Columbia), 111 p.

KING, F. H. [1899], Movement of ground water; *Nineteenth Annual Report of the United States Geological Survey*, Part 2, 59–294.

MARTEL, E. A. [1894], *Les Abîmes* (Paris), 580 p.

LAVALLE, P. [1967], Some aspects of linear karst depression development in south central Kentucky; *Annals of the Association of American Geographers*, **57**, 49–71.

OLLIER, C. D. [1960], The inselbergs of Uganda; *Zeitschrift für Geomorphologie*, **4**, 43–52.

PITTY, A. F. [1968], The scale and significance of solutional loss from the limestone tract of the southern Pennines; *Proceedings of the Geologists' Association*, **79**, 153–77.

PRESCOTT, J. A. and PENDLETON, R. L. [1952], *Laterite and Lateritic Soils;* Commonwealth Bureau of Soil Science (Rothamsted), Technical Communication No. 47, 51 p.

RHOADES, R. and SINACORI, M. N. [1941], Pattern of ground-water flow and solution; *Journal of Geology*, **49**, 785–94.

SWINNERTON, A. C. [1932], Origin of limestone caverns; *Bulletin of the Geological Society of America*, **43**, 663–93.

THOMAS, M. F. [1966], Some geomorphological implications of deep weathering patterns in crystalline rocks in Nigeria; *Transactions and Papers of the Institute of British Geographers*, **40**, 173–93.

WARWICK, G. T. [1962], The origin of limestone caverns; in Cullingford, C.H.D., Editor, *British Caving*, (2nd Edition), (London), pp. 55–85.

WILLIAMS, P. W. [1966], Morphometric analysis of temperate karst landforms; *Irish Speleology*, **1**, 23–31.

WILLIAMS, P. W. [1968], An evaluation of the rate and distribution of limestone solution and deposition in the river Fergus basin, western Ireland; *Australian National University, Research School of Pacific Studies, Department of Geography, Publication* G/5, 1–40.

WOOLNOUGH, W. G. [1927], The duricrust of Australia; *Journal and Proceedings of the Royal Society of New South Wales*, **61**, 24–53.

WOPFNER, H. and TWIDALE, C. R. [1967], Geomorphological history of the Lake Eyre basin; in Jennings, J. N. and Mabbutt, J. A., Editors, *Landform Studies from Australia and New Guinea* (Canberra), pp. 118–43.

ZÖTL, J. [1965], Tasks and results of karst hydrology; in Štelcl, O., Editor, *Problems of the Speleological Research*, (Prague), pp. 141–5.

7.I. Open Channel Flow

D. B. SIMONS

Department of Civil Engineering, Colorado State University

1. Introduction

Flow in open channels has been nature's way of conveying water on the surface of the earth since the beginning of time. Furthermore, these streams have constantly been the subject of study by man since he has been alternately blessed by the life-giving quality of streams under control and plagued by their destructive quality when out of control, such as in time of flood. Hence, the characteristics of rivers are of importance to everyone dealing with water resources, whether from the viewpoint of geomorphology, hydraulics, flood control, navigation, stabilization or water-resources development for municipalities, and industry.

2. Properties of fluids

The following physical properties of fluids influence fluid motion, channel geometry, and help explain sediment transport.

Mass is the amount of substance in matter measured by its resistance to the application of force.

Density is mass per unit volume, and is commonly symbolized by the Greek letter ρ (rho).

Weight is the force that gravity exerts on a mass; $W = gM$, where g is the acceleration of gravity.

Specific Weight is the weight per unit volume and is symbolized by the Greek letter γ (gamma); $\gamma = \rho g$.

Viscosity is the property of fluids that resists deformation and is commonly symbolized by the Greek letter μ (mu).

Shear is a property of fluid motion that is closely related to viscosity. It is the tangential force or stress per unit area that is transmitted through a unit thickness of a fluid. Shear, τ (tau), is related to viscosity by the equation $\tau = \mu \dfrac{dv}{dy}$.

Temperature affects the density of liquids slightly and the viscosity significantly. That is, water is essentially incompressible. The viscosity of liquids decrease with increasing temperature. Water temperature in open channels can vary as much as 40° F within a 24-hour period.

Elasticity and *Surface Tension* have little effect on flow in open channels, including sediment transport.

3. Types of flow

There are several types of flow in open channels. These include laminar flow and turbulent flow; uniform flow and non-uniform (or varied) flow; steady flow and unsteady flow; and tranquil flow, rapid flow, and ultra-rapid flow.

A. Laminar flow

Fluid motion may occur as laminar or turbulent flow. In laminar flow each fluid element moves along a specific path with a uniform velocity. There is no diffusion between the stream tubes, layers, or elements of flow; and accordingly, there is no turbulence. The energy used in maintaining viscous flow is dissipated in the form of heat from the friction within the fluid.

With laminar flow, the shear stress $\tau = \mu \dfrac{dv}{dy}$ being transmitted through each unit of depth varies uniformly from zero at the surface to a maximum at the stream bed, while the velocity curve is parabolic in shape, with its vertex at the surface.

In stream flow disturbances are present in such magnitude that laminar flow is rarely found. As velocity or depth increases, a given condition of laminar flow will reach a critical condition and become turbulent flow.

B. Reynolds' number

Values of *Reynolds' number* (*Re*) can be used to predict the type of flow. This dimensionless number includes the effects of the flow characteristics, velocity, and depth, and the fluid properties density and viscosity.

$$Re = \frac{VR\rho}{\mu} \tag{1}$$

The ratio $\dfrac{\mu}{\rho}$ is a fluid property called the kinematic viscosity, commonly designated ν (nu). Using this property,

$$Re = \frac{VR}{\nu} \tag{2}$$

With the value of Reynolds' number less than 500, laminar flow will prevail; whereas, with values in excess of 750, turbulent flow will prevail for smooth boundary conditions. For natural channels the critical value will be near 500 due to bed roughness.

The Reynolds' number is defined as

$$Re = \frac{\text{Inertia force}}{\text{Viscous force}} \tag{3}$$

That is, *Re* is an index of the relative importance of viscous forces in a hydraulic problem. Using Newton's second law of motion to define the inertial force, the expression $\tau = \mu \dfrac{dv}{dy}$ to define the viscous force and dimensional analysis

$$Re = \frac{\rho\dfrac{L^3L}{T^2}}{\mu\dfrac{LL^2}{TL}} \qquad (4)$$

substituting

$$V = \frac{L}{T} \quad \text{and} \quad L = D$$

$$Re = \frac{\rho V^2 D^2}{\mu VD} = \frac{VD\rho}{\mu} \qquad (5)$$

Many other dimensionless parameters which have the same form as the foregoing Reynolds' number are utilized in the analysis of open-channel flow problems, such as: $\dfrac{wd}{\nu}$ and $\dfrac{V_* d}{\nu}$,

where w = fall velocity of sediment or bed material;

d = median diameter of the sediment or bed material;

V_* = shear velocity which is equal to $\sqrt{gRS}$.

Problem No. 1. A sheet of water 0·25 ft deep is flowing over a smooth surface at 1·0 ft/sec. Compute the Reynolds' number (Re) of the flow. Will the flow likely be laminar or turbulent? (Kinematic viscosity (μ) = 1·21 × 10^{-5} ft^2/sec).

$$Re = \frac{(1\cdot0)\,(0\cdot25)}{1\cdot21}\,(10^5) = 20{,}700 \text{ Turbulent}$$

C. Froude number

The Froude number F_r is another dimensionless parameter frequently used to describe flow conditions. It is an index to the influence of gravity in flow situations where there is a liquid–gas interforce – such as in an open channel. The Froude number is usually defined as

$$F_r = \left(\frac{\text{Inertia force}}{\text{Gravity force}}\right)^{\frac{1}{2}} \qquad (6)$$

or

$$F_r = \frac{\text{Velocity of flow}}{\text{Velocity of a small gravity wave in still water}} \qquad (7)$$

Referring to the first definition and dimensional analysis

$$F_r = \left(\frac{\rho\dfrac{L^3L}{T^2}}{\Delta\gamma L^3}\right)^{\frac{1}{2}} \qquad (8)$$

Substituting

$$V = \frac{L}{T}$$

$$F_r = \left(\frac{\rho V^2 L^2}{\Delta\gamma L^3}\right)^{\frac{1}{4}} = \frac{V}{\sqrt{\Delta\gamma L/\rho}} \tag{9}$$

where L = a length dimension;

$\Delta\gamma$ = difference in specific weight of the fluids -- usually air and water;

V = average velocity of flow;

ρ = mass density of the fluid.

In open-channel flow $\Delta\gamma$ is essentially the same as γ for water alone, since the density of air is so small. If D (depth) or hydraulic radius $R = \frac{A}{P}$ is used for the L dimension and $\frac{\gamma}{\rho}$ replaced by its equivalent g, then the Froude number becomes

$$F_r = \frac{V}{\sqrt{Dg}} \tag{10}$$

Note that in wide shallow channels depth of flow and the hydraulic radius are nearly equal.

A Froude number of unity indicates critical flow; less than unity indicates tranquil flow, the common variety of turbulent flow; and greater than unity, 'rapid flow'. In keeping with the second definition of Fr the $\sqrt{gD}$ term is the velocity at which a small wave travels in still water of depth D. For example, one can throw a pebble into a stream and, comparing the velocity of the waves caused by the pebble and the velocity of the flow, determine whether flow is sub-critical, critical, or super-critical.

Problem No. 2. Compute the Froude number of an open-channel flow where the mean velocity is 5 ft/sec and the depth is 1·5 ft. Describe the flow with respect to critical flow.

$$F_r = \frac{V}{\sqrt{gD}}$$

$$F_r = \frac{5}{\sqrt{(32\cdot2)(1\cdot5)}} = 0\cdot73 \text{ tranquil or subcritical}$$

D. Turbulent flow

Turbulence, as a complicated pattern of eddies, produces small velocity fluctuations at random in all directions with an average time value of zero. Energy dissipation is high in turbulent flow due to the continuous interchange of finite masses of fluid between neighbouring zones of flow. The resistance to flow increases with approximately the square of the velocity.

Turbulent flow, as a result of this mixing and exchange of energy, has a more uniform distribution of velocity from top to bottom than laminar flow. The velocity distributions for laminar and turbulent flow are compared qualitatively in fig. 7.1.1.

Several parameters have been developed to describe turbulent flow. The shear stress in turbulent flow is defined as $\tau = \eta \dfrac{dv}{dy}$, in which η (eta) is termed the *eddy viscosity*. A parameter used to describe the magnitude of turbulent velocity fluctuations is the root-mean-square $(\sqrt{V'^2})$ of the deviations from the mean

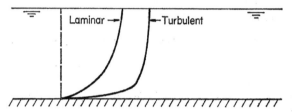

Fig. 7.1.1 Comparison of velocity distribution in laminar and turbulent flow.

velocity. The mean size of the turbulent eddies is measured by the mixing length (l) – the distance through which fluid elements move before diffusing with the surrounding fluid. The diffusion coefficient, $\epsilon = l\sqrt{V'^3}$, is a measure of the mixing process. The general pattern of variation of these parameters in turbulent flow is shown in fig. 7.1.2.

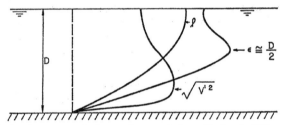

Fig. 7.1.2 Variation in mean turbulence characteristics with depth (After Rouse, 1946).

From a combination of experimental study and theory the distribution of velocity in turbulent flow over smooth boundaries has been determined. A thin layer of laminar flow persists at the boundary surfaces. This layer is called the laminar sub-layer. The theoretical velocity curve is a composite of the logarithmic turbulent flow pattern and the nearly linear laminar pattern joined by a transition curve as shown in fig. 7.1.3.

In fig. 7.1.3 δ' is the thickness of the laminar sub-layer, and it is defined by the equation

$$\delta' = \frac{11 \cdot 6\, \nu}{\sqrt{\dfrac{\tau}{\rho}}} = \frac{11 \cdot 6\, \nu}{V_*} \tag{11}$$

If extended towards the bed the logarithmic form of the turbulent velocity distribution will yield zero velocity at a distance y' above the bed. Experiments show that $y' = \delta'/107$.

The term $\sqrt{\dfrac{\tau}{\rho}}$ is a common parameter called the shear velocity (V_*), and is equal to $\sqrt{RSg}$.

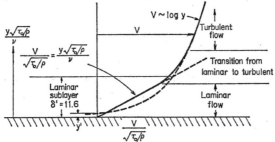

Fig. 7.1.3 Details of flow near the bed of an open channel (After Albertson, Barton and Simons, 1960).

The Karman–Prandtl equation describes the velocity distribution of turbulent flow over a smooth bed as

$$\frac{v}{V_*} = 5 \cdot 75 \log_{10} \frac{V_* y}{\nu} + 5 \cdot 5 \text{ for the turbulent zone}$$

and

$$\frac{v}{V_*} = \frac{V_* y}{\nu} \text{ for the laminar zone}$$

The two equations are presumed to be joined by a smooth transition curve at the distance δ' above the bed.

Uniform flow in open channels depends on there being no change with distance in either the magnitude or direction of the velocity along a stream line, i.e. both $\partial v/\partial s = 0$ and $\partial v/\partial n = 0$. Non-uniform flow in open channels occurs when either $\partial v/\partial s \neq 0$ or $\partial v/\partial n \neq 0$. Varied flow in open channels is a type of non-uniform flow which occurs when $\partial v/\partial s \neq 0$. Steady flow occurs when the velocity at a point does not change with time, i.e., $\partial v/\partial t = 0$. When the flow is unsteady $\partial v/\partial t \neq 0$. An example of unsteady flow is a flood wave or a travelling surge.

Unlike laminar and turbulent flow, tranquil flow and rapid flow exist only with a free surface or inner face. The criterion for tranquil and rapid flow is the Froude number $F_r = V/(gD)^{1/2}$. When $F_r = 1 \cdot 0$ the flow is critical; when $F_r < 1 \cdot 0$ the flow is tranquil; and when $F_r > 1 \cdot 0$ the flow is rapid. Ultra-rapid flow involves slugs or waves superposed over the uniform flow pattern, which makes the flow both non-uniform and unsteady.

Uniform flow in an open channel occurs with either a mild, a critical, or a steep slope. With a mild slope the flow is tranquil; with a critical slope the flow is critical; and with a steep slope the flow is rapid.

4. Velocity distribution over rough beds

Stream channels have rough beds. The roughness is expressed in terms of K_s, which is equivalent to the diameter of the sediment grains which compose the bed. The dimension of K_s is larger than that of δ', and therefore the sub-layer ceases to exist for practical purposes. Turbulent flow is assumed to occur throughout the depth. The Karman–Prandtl velocity equation for rough beds is

$$\frac{v}{V_*} = \frac{2 \cdot 303}{\kappa} \log_{10} \frac{y}{K_s} + 8 \cdot 5 \qquad (12)$$

where kappa (κ) is the so-called universal velocity coefficient, which is approximately 0·4 for fixed boundary channels. The distribution of velocity in accordance with this equation is a straight line when plotted on semi-log paper with a slope of $\dfrac{\kappa}{2 \cdot 303\ V_*}$.

5. Velocity and discharge measurements

Velocity is a vector quantity, hence both its direction and magnitude must be measured. The discharge in an open channel or pipe, in the most simple terms, is the product of area and average velocity measured normal to the area.

A. Current meters

In an open channel one of the most common methods of measuring discharge involves integration of the velocity distribution across the flow section using a current meter, pitot tube, or similar device. The current meter consists of an instrument with an impeller mounted on a rod or cable. If a cable is used to suspend the meter in the flow there must be a streamlined weight at its lower end, below the current meter, of sufficient magnitude to overcome the force of the stream, enabling the operator to place the meter at any desired point in the vertical. Having taken sufficient point measurements in a vertical to establish the average velocity, the operator moves to a new vertical, by wading in small streams or by cable car or boat on large streams.

The average velocity in a vertical is located at approximately 0·6 depth below the surface and can be more precisely determined by averaging the point velocities at 0·2 and 0·8 depth. In flows with a depth less than 1·5 ft a single-point measurement is taken in each vertical in the stream cross-section at 0·6 depth. In deeper streams the 0·2 and 0·8 measurements are taken and averaged in each vertical. Discharge determined by the 0·2 and 0·8 measurements is illustrated in the example problem.

Problem No. 3. The computation of stream discharge based on current meter measurements is illustrated in Table 7.1.1.

Similarly, floats can be used to estimate the magnitude and direction of surface velocities. Greater accuracy can be achieved by using a submerged or partly submerged float which measures the velocity at more nearly 0·6 depth. Also, it is more independent of the effect of wind and waves, but may be bothered by debris.

TABLE 7.I.I

Distance from bank	Depth	Observation depth	Velocity At Point	Velocity Mean in vertical	Velocity Mean in section	Area	Mean depth	Width	Discharge (q)
0	0	0	0	0					
					0·78	1·70	0·85	2	1·33
2	1·70	0·35	1·52	1·56					
		1·35	1·60						
					1·73	4·40	2·20	2	7·61
4	2·70	0·54	1·91	1·89					
		2·16	1·88						
					2·08	13·80	3·45	4	28·7
8	4·20	0·84	2·35	2·27					
		3·36	2·19						
					2·33	8·60	4·30	2	20·1
10	4·40	0·88	2·41	2·38					
		3·52	2·34						
					2·37	8·40	4·20	2	19·94
12	4·00	0·80	2·31	2·36					
		3·20	2·40						
					2·05	13·60	3·40	4	27·9
16	2·80	0·56	1·92	1·74					
		2·24	1·57						
					1·49	4·30	2·15	2	6·41
18	1·50	0·30	1·35	1·24					
		1·20	1·13						
					0·62	1·50	0·75	2	0·93
20	0	0	0	0					

$$Q = \Sigma q = 112\cdot9 \text{ cfs.}$$

B. Dye-dilution

Various fluorescent tracer-type dyes can be detected and accurately measured at very low concentrations using a fluorometer. This makes it possible to successfully measure discharge by various dye-dilution techniques. Two methods can be used. One involves

$$Q = q\frac{C_1}{C_2} \tag{13}$$

where q is the injected discharge, C_1 is concentration of the dye in the injected flow, and C_2 is the concentration of dye in the unknown discharge Q. On small streams injection of a small steady q at known concentration C_1 for about 15 minutes will enable C_2 to stabilize. Only C_2 at its plateau needs to be determined to compute the stream discharge Q since the effect of q is small.

The second method is called the total recovery method. The discharge is evaluated using the relation

$$Q = \frac{(\text{Vol. of dye})(\text{Conc. of dye})}{\int_0^\infty C \, dt} \tag{14}$$

The integral term is the total area under the concentration–time curve, where C is the measured dye concentration at time t at the point of sampling. In both relations any material background fluorescence must be considered in measuring effective dy concentrations.

C. Weirs

The weir is extensively used to measure flow in open channels. It is essentially an overflow structure extending across the channel normal to the direction of flow, see fig. 7.1.4.

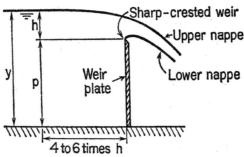

Fig. 7.1.4 Sharp-crested weir.

Weirs are classified according to shape. The most common ones are the standard uncontracted weir, also known as the suppressed weir, the contracted weir, the V-notch weir, the trapezoidal weir, and the broad-crested weir. The first four are sharp crested, as shown in fig. 7.1.4.

Many formulae have been suggested by various experimenters, but only a few are presented.

In 1823 Francis suggested the equation

$$Q = 3 \cdot 33 L h^{3/2} \tag{15}$$

for uncontracted (suppressed) weirs in which L is the length of crest in feet and h is the head on the weir in feet. For the contracted weir

$$Q = 3 \cdot 33 \left(L - \frac{nh}{10} \right) h^{3/2} \tag{16}$$

where n is the number of horizontal end contractors with a simple weir only contracted at the sides of the channel $n = 2$.

For a triangular weir $\quad dQ = C_d x \sqrt{2gy}\, dy$ (17)

from which $\quad Q = \dfrac{8}{15} C_d \sqrt{2g}\, \tan\dfrac{\theta}{2}\, h^{5/2}$ (18)

see fig. 7.1.5.

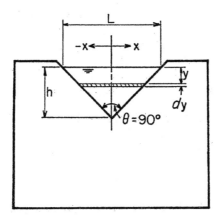

Fig. 7.1.5 Triangular weir.

The weirs should be installed so that:

1. The weir plate is vertical and the upstream face essentially smooth.
2. The crest is horizontal and normal to the direction of flow. The crest must be sharp, so that the water springs free from the edge.
3. The pressure along the upper and the lower nappe is atmospheric.
4. The approach channel is uniform in cross-section, and the water surface is free of surface waves.
5. The sides of the channel are vertical and smooth, and they extend downstream from the crest of the weir.

in order for the equations to apply.

Still greater accuracy can be achieved with all types of measuring devices if they are calibrated after construction.

Many other methods of measuring discharge in open channels and pipes are utilized. The principles utilized are (1) volumetric, (2) use of the Bernoulli momentum and continuity equations, (3) drag on an object in the flow, (4) mass flux measurements, and (5) by the slope area method. The measuring devices include: Venturi meters, Parshall flumes, nozzles, orifices, gates, weirs, spillways, contracted openings, vane meters, rotometers, and wobbling discs. Refer to any fluid mechanics, such as Albertson, Barton, and Simons [1960], for further details.

6. Hydraulic and energy gradients

The hydraulic gradient in open-channel flow is the water surface. The energy gradient is above the hydraulic gradient a distance equal to the velocity head.

The fall of the energy gradient for a given length of channel represents the loss of energy, either from friction or from friction and other influences. The relationship of the energy gradient to the hydraulic gradient reflects not only the loss of energy but also the conversion between potential and kinetic energy. For uniform flow the gradients are parallel and the slope of the water surface represents the friction-loss gradient. In accelerated flow the hydraulic gradient is steeper than the energy gradient, indicating a progressive conversion from potential to kinetic energy. In retarded flow the energy gradient is steeper than the hydraulic gradient, indicating a conversion from kinetic to potential energy. The Bernoulli theorem defines the progressive relationships of these energy gradients.

For a given reach of channel ΔL, the average slope of the energy gradient is $\Delta h_L/\Delta L$, where Δh_L is the cumulative loss through the reach. If these losses are solely from friction, Δh_L will become Δh_f and

$$\Delta h_f = \frac{S_2 + S_1}{2} \Delta L \qquad (19)$$

where S_1 and S_2 are the slopes of the energy gradient at the ends of reach ΔL.

7. Energy and head

If streamlines of flow in an open channel are parallel and velocities at all points in a cross-section are equal to the mean velocity V the energy possessed by the water is made up of kinetic energy and potential energy. Referring to fig. 7.1.6,

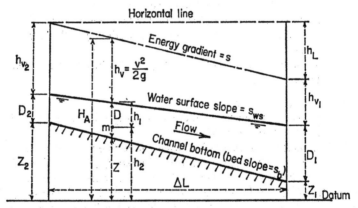

Fig. 7.1.6 Characteristics of open-channel flow.

the potential energy of mass M is $\gamma(h_1 + h_2)$ and the kinetic energy of M is $\gamma \frac{V^2}{2g}$. Hence, the total energy of each mass particle is:

$$E_m = \gamma h_1 + h_2 + \frac{V^2}{2g} \qquad (20)$$

Applying the above relationship to the total discharge Q in terms of the unit weight of water γ,

$$E = Q\gamma \left(D + Z + \frac{V^2}{2g} \right) \tag{21}$$

where E is total energy per second at the cross-section.

The parentheses term in equation (21) is the absolute head:

$$H_A = D + Z + \frac{V^2}{2g} \tag{22}$$

Equation (22) is the Bernoulli Equation.

The energy in the cross-section referred to the bottom of the channel is termed the specific energy. The corresponding head is referred to as the specific energy head, and is expressed as:

$$H_E = D + \frac{V^2}{2g} \tag{23}$$

Where $Q = AV$, equation (23) can be stated:

$$H_E = D + \frac{Q^2}{2ga^2} \tag{24}$$

8. Flow equations

One of the common open-channel flow equations for estimating average velocity is the Chezy equation

$$V = (8g/f)^{1/2}(RS)^{1/2} = C(RS)^{1/2} = C/g^{1/2}(gRS)^{1/2} \tag{25}$$

Bazin [1897] suggested that

$$C = 157 \cdot 6/[1 + (k_1/R^{1/2})] \tag{26}$$

where k_1 is the roughness coefficient varying from 0·11 for very smooth cement or planed wood to 3·17 for earth channels in rough condition. In this equation the upper limit for C is 157·6.

In 1911 Johnston and Goodrich proposed using an exponential formula of the form,

$$V = CR^p s^q \tag{27}$$

and gave values of C and p, making q uniformly equal to 0·5 for simplicity (Ellis, 1916). This is exactly the same formula as proposed by Chezy, where the numerical value of the Chezy's coefficient C is equal to 0·5. Other open-channel flow equations that are often used are given by N. G. Bhowmik [1965] and Garbrecht [1961].

In an effort to correlate and systematize existing data from natural and artificial channels, Manning [1889] proposed the equation

$$V = (1 \cdot 5/n)R^{2/3} S^{1/2} \tag{28}$$

or

$$Q = AV = A(1 \cdot 5/n)R^{2/3} S^{1/2} \tag{29}$$

in which n is the Manning roughness coefficient which has the dimensions of $L^{1/6}$. By comparing equation (25) with equation (28), the Chezy discharge coefficient C can be expressed as follows:

$$C = 1 \cdot 5 (R^{1/6}/n) \tag{30}$$

and is related to the Manning coefficient n and the hydraulic radius $R = \dfrac{A}{P}$.

The Manning n was developed empirically as a coefficient which remained a constant for a given boundary condition, regardless of slope of channel, size of channel, or depth of flow. As a matter of fact, however, each of these factors causes n to vary to some extent. In other words, the Reynolds' number, the shape of the channel, and the relative roughness have an influence on the magnitude of Manning's n. Furthermore, for a given alluvial bed of an open channel the size, pattern, and spacing of the sand waves vary, so that n varies. Despite the shortcomings of the Manning roughness coefficient, it is used extensively.

The magnitude of Manning roughness is given in Table 7.1.2 for rigid channels

TABLE 7.1.2 Manning roughness coefficients for various boundaries

Boundary	Manning Roughness n in $(ft)^{1/6}$
Very smooth surfaces such as glass, plastic, or brass	0·010
Very smooth concrete and planed timber	0·011
Smooth concrete	0·012
Ordinary concrete lining	0·013
Good wood	0·014
Vitrified clay	0·015
Shot concrete, untrowelled, and earth channels in best condition	0·017
Straight unlined earth canals in good condition	0·020
Rivers and earth canals in fair condition – some growth	0·025
Winding natural streams and canals in poor condition – considerable moss growth	0·035
Mountain streams with rocky beds and rivers with variable sections and some vegetation along banks	0·040–0·050
Alluvial channels, sand bed, no vegetation	
1. Tranquil flow, $F_r < 1$	
Plane bed	0·014–0·02
Ripples	0·018–0·028
Dunes	0·018–0·035
Washed-out dunes or transition	0·014–0·024
Plane bed	0·012–0·015
2. Rapid flow $F_r > 1$	
Standing waves	0·011–0·015
Anti-dunes	0·012–0·020

and alluvial channels. Considering alluvial channels, note that as the form of bed roughness changes from dunes through transition to plane bed or standing waves the magnitude of Manning n decreases by approximately 50%.

9. Natural channels

The natural shape of an open channel may be markedly different from rectangles and trapezoids. However, it is usually possible to break down the complex shape of a natural open channel into simple elementary shapes for analysis. Consider fig. 7.1.7, in which flow is occurring not only in the main channel but also in the

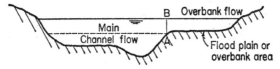

Fig. 7.1.7 Shape of natural channels.

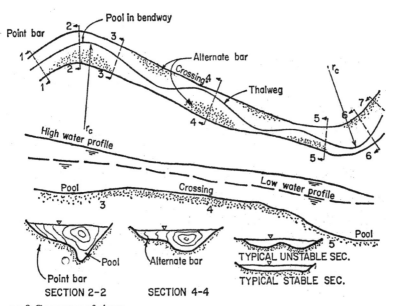

Fig. 7.1.8 Geometry of rivers.

overbank or floodplain area. In this case the hydraulic radius R, which would be obtained by using the area and the wetted perimeter for the entire section, would not be truly representative of the flow. Furthermore, the roughness coefficient in the overbank area is usually different from the coefficient in the main channel. Therefore, such a section should be divided along AB and treated as two separate sections. The plane AB, however, is not considered as a part of the wetted perimenter, since there is no appreciable shear in this plane.

Along a natural channel there are frequently pools and riffles or rapids. At low to moderate discharges the slope of water surface is relatively flat over the pools and steep over the riffles. With a further increase in stage, this condition may be reversed as for the Mississippi River (fig. 7.1.8). Therefore care must be taken in studies of natural streams to consider the correct slope for the particular discharge and reach of stream in question.

10. Forms of bed roughness and resistance to flow in alluvial channels

The primary variables which affect the form of bed roughness and resistance to flow in sand-bed alluvial channels (Simons and Richardson, 1962, 1963) include:

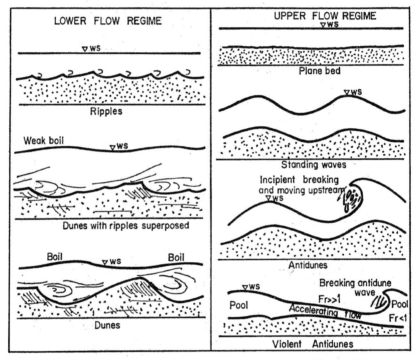

Fig. 7.1.9 Forms of bed roughness in alluvial channels (the term 'flat bed' is now preferred to 'plane bed').

the slope of the energy grade line, depth, physical size of the bed material as related to grain roughness, and fall velocity or effective median fall diameter as related to form resistance. The fall velocity or effective median fall diameter depends on the viscosity and mass density of the water sediment mixture and the mass density, size, and shape of the bed material. It reflects the principal viscous effect on flow in alluvial channels when Re is large. The effective median fall diameter is defined as the diameter of a sphere having a specific gravity of 2·65 and a fall velocity in distilled water of infinite extent at a temperature of

24° C equal to the fall velocity of the particle falling alone in any quiescent stream fluid at stream temperature (Haushild *et al.*, 1961; Simons *et al.*, 1963).

The regimes of flow and various forms of boundary roughness (Simons and Richardson, 1963) which can occur in alluvial channels are illustrated in fig. 7.1.9.

In the lower flow regime flow is tranquil and the water surface undulations are out of phase with the bed undulations. Resistance to flow is large, because separation of the flow from the boundary generates large-scale turbulence that dissipates considerable energy.

With depths of flow ranging from 0·4 to 1·0 ft, ripple heights range from 0·01 to 0·1 ft and length (crest-to-crest) range from 0·5 to 1·5 ft. When depth is small the ripples increase in size with depth, but at greater depths ripple size becomes independent of depth. Therefore, ripples observed in flumes are similar in size and shape to those in natural streams. A decrease in n occurs when depth is increased, indicating a relative roughness effect or when effective fall diameter is increased, which causes a decrease in ripple size. The decrease in n with an increase in effective fall diameter is similar to change reported in Leopold and Maddock [1953]. Apparently, ripples do not form when the median diameter of the bed material is coarser than 0·7 mm.

With depth of flow ranging from 0·4 to 1·0 ft, dune heights range from 0·15 to 1·0 ft and dune length from 4 to 20 ft, based upon flume studies (Simons and Richardson, 1963). In deep rivers dunes 30–60 ft in height and with lengths of several hundred feet have been observed. In the flume studies n increased with depth because size of the dunes and, hence, scale and intensity of turbulence increased. This may not be the case as larger depths are studied. With an increase in slope, n decreased for the fine sand but increased for the coarse sand because dune length increased appreciably with the fine sand but did not with the coarse sand. An increase in effective fall diameter increased n because dune length decreased and dune angularity increased. The long dunes formed by the finer sands exhibited smaller n values than ripples.

In the transition zone n varied from the largest value for the lower flow regime to the smallest value for the upper flow regime. In this zone a well-defined relation between n and boundary shear does not exist. The bed form in the transition zone depended, in addition to the other factors, on antecedent conditions. Starting with dunes, slope and/or depth could be increased to relatively large values before plane bed or standing waves occurred. Conversely, with a plane bed and/or standing waves, the slope and/or depth could be decreased to relatively small values before dunes developed.

In the upper flow regime n values are small because surface or grain resistance predominates. However, the energy dissipated by the wave formation with the standing waves and the formation and breaking of the waves with antidunes increases n. Standing waves and antidunes, from the standpoint of wave mechanics, are rapid flow phenomena.

Standing waves are sinusoidal, in-phase sand and water waves (fig. 7.1.9) that build up in amplitude from a plane bed and water surface and gradually fade away. In the flume studies with depth of flow ranging from 0·2 to 0·6 ft the water wave height (trough-to-crest) ranged from 0·01 to 0·6 ft and was 1·5–2 times the height of corresponding sand waves. Spacing of the waves was from 2 to 5 ft. Both height and spacing of the waves increased with depth. Resistance to flow for standing waves was larger than for a plane bed and, as with a plane bed, increased with an increase in sand size. Standing waves did not occur using the two finer sands, because the mobility of the particles (effective fall diameter) allowed the development of antidunes whenever the Froude number equalled one.

Antidunes are similar to standing waves, except they increase in amplitude until they break. Breaking antidunes are similar to the hydraulic jump. The breaking wave dissipates a large amount of energy that is reflected by increased n. The increase of n is in direct proportion to the amount of antidune activity and the portion of the flume or channel occupied by the antidunes. Antidune activity increases with a decrease in effective fall diameter or with an increase in slope.

11. Comparison of flume and field phenomena

The preceding comments were based on observations of bed configurations and flow phenomena that occurred in the flume experiments and in natural rivers (field conditions) and are equally true for both situations. However, there are major differences between flume and field conditions. In the usual flumes only a limited range of depth and discharge can be investigated, but slope and velocity can be varied within a wide range. In the field a larger range of depth and discharge is common, but slope of a particular channel reach is virtually constant. Larger Froude numbers $(V/\sqrt{gD})$ can be achieved in flume studies than will occur in most natural alluvial channels because natural banks cannot withstand prolonged high-velocity flow without eroding. This erosion increases the cross-sectional area, and this causes a reduction in the average velocity and the Froude number. Rarely does a Froude number, based on average velocity and depth, exceed unity for any extended time period in a natural stream with erodible banks. In fact, rarely are natural channels truly stable when $F_r > 0·25$. In the field, where the slope of the energy grade line is constant, the Froude number is also constant unless there is a change in the resistance to flow,

$$\left(F_r = \frac{C}{\sqrt{g}}\sqrt{S}\right)$$

Flow is more nearly two-dimensional in flumes than in natural streams. However, the main current meanders from side to side in a large flume, as it does in the field, and bars of small amplitude but large area develop in an alternating pattern adjacent to the walls of the flume. (It is on these bars that the bed forms shown in fig. 7.1.9 superpose themselves.) If very large width–depth ratios are maintained by keeping depth of flow shallow these bars may grow to the water surface.

In the field it is even more obvious that the flow meanders between the parallel banks of a straight channel, and the alternate bars which form opposite the apex of the meanders are easier to distinguish. As in a flume, if the depth is decreased the alternate bars increase in amplitude until they are close to the water surface, or even exposed. In fact, scour in the main current adjacent to a large bar may cause the water surface there to drop slightly so that the top of the bar is exposed. This has been observed in the Rio Grande and in other natural channels. If the banks are not stable erosion occurs where the high-velocity water impinges, and deposition occurs on the opposite bank. The ultimate development is a meandering stream if other factors such as slope, discharge, and size of bed material are compatible.

The alluvial bars which form on the bed of an alluvial channel is a type of roughness element that plays a very significant role in river mechanics and channel geometry. These bars are much larger than the ripples and dunes illustrated in fig. 7.1.9. Their amplitude ranges from very small to as large as the average depth of flow. Their widths range from $\frac{1}{2}$ to nearly $\frac{1}{3}$ channel width, depending upon the size of the system, and they may be several hundred feet long.

The three most common and accepted types of bars are: point bars, alternate bars, and middle bars (see fig. 7.1.8).

The position, shape, and magnitude of the alternate bars is a function of channel alignment, bed material, and width–depth ratio. These bars normally occur in the straight reaches or crossings between two consecutive bends. Conditions in the upstream bend usually dictate on which side of the channel the first alternate bar will form. Thereafter the sequence is fixed, because the second bar must be on the opposite side of the channel, and so on. In the next bend downstream the secondary circulation developed within it is usually, unless the radius of the bend is very large, of sufficient magnitude to terminate the sequence of alternate bars. Within each bend the pool forms adjacent to the outside bank and a point bar forms on the inside bank. Consequently, the alternate bars are in a sense locked in between the two bends, and their physical characteristics vary with the type of bed material, the width–depth ratio of the channel, and the characteristics of the bends. The number of bars between two bends may increase as the discharge decreases, and vice versa. The flow meanders around these alternate bars. In fact, this flow phenomena is probably the way meanders are initiated and develop with time. For example, consider the successive stages of the development of a meander in sand. Initially the channel may be straight, but the meandering of the thalweg rapidly initiates the development of the meander. Another significant point is that the wavelength of the meander is essentially constant for constant discharge throughout the development of the meander, even though the amplitude increases significantly. The rate of movement of the alternate bars is mostly just a change in size and shape as flow conditions change and the bars alter their geometry to suit the new flow conditions. However, if the bends of the channel are moving, then this allows an additional freedom of movement for the bars. Leopold [1964] has documented that bars move very slowly – of the order of a foot or so per day. Simons has verified the foregoing information on bars by observing their development and movement with time in canals and rivers, and has studied their detailed movement in the sand-bed flumes at Colorado State University. He observed that as a specific weight of the bed material decreases the rate of both change of shape and movement downstream rapidly increases.

Various methods for predicting form of bed roughness have been developed. One of the more useful methods was presented by Simons and Richardson [1963].

12. Tractive force

The tractive-force theory is formulated on the basis that stability of bank and bed material is a function of the ability of the bank and bed to resist erosion resulting from the tractive force exerted on them by the moving water.

Consider the free body of a segment of the full width of the channel as shown in fig. 7.1.10.

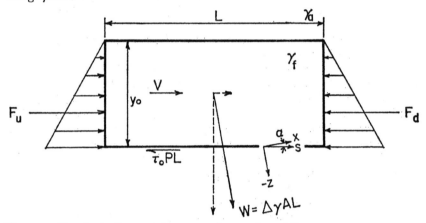

Fig. 7.1.10 Free-body diagram of segment of open-channel flow.

Equating the forces parallel to the flow yields

$$F_u + W \sin \alpha = F_d + \tau_0 p L \tag{31}$$

where W is the weight of the entire segment of fluid;
   p is the wetted perimeter – that is, the length of the cross-sectional boundary which is in contact with the fluid flowing in the channel;
   F_u and F_d are the upstream and downstream hydrostatic forces acting on the free body where $p = \gamma y$ – since the flow is uniform, $F_u = F_d$;
   τ_0 is the average boundary shear which is retarding the flow;
   A is the cross-sectional area of the flow;
   L is the length of the free body segment;
   α is the angle which the channel slope makes with the horizontal; and
   $\Delta\gamma$ is the difference between the specific weight γ of the fluid flowing and the specific weight γ_a of the ambient fluid, normally the air.

The product $\Delta \gamma A L$ may be substituted for the weight W and the equation rearranged to solve for the boundary shear

$$\tau_0 = \Delta \gamma \frac{A}{wp} \sin \alpha - \Delta \gamma \, RS = \gamma \, RS \tag{32}$$

where R is called the hydraulic radius which is the area A divided by the wetted
    perimeter p; and
    S is $\sin \alpha = dz/ds$, which for relatively flat slopes may be considered more
    conveniently as $\tan \alpha = dz/dx$.

Equation (31), it may be noted, evaluates the boundary shear in terms of the static characteristics of the geometry and the fluid.

A tractive force theory was clearly presented and illustrated by Lane [1953] to assist with the design of channels for conveying essentially clear water in coarse non-cohesive materials and where bank stabilization is to be achieved by armour plating with coarse non-cohesive material. For a review of the design of alluvial channels in accordance with regime and other concepts, refer to Lacey [1958] and Simons and Albertson [1963].

13. Sediment transport

Knowledge of sediment transport in alluvial channels is just as important as knowledge of resistance to flow. The ability of a stream to transport bed material

TABLE 7.1.3 Variation of concentration on a dry-weight basis of total bed
material load with regimes of flow and forms of bed roughness

		Total bed material load (p.p.m.)	
Regime of flow	Forms of bed roughness	Median diameter of bed material 0·28 mm	Median diameter of bed material 0·45 mm
Lower flow regime	Ripples	1–150	1–100
	Dunes	150–800	100–1,200
Transition	Zone in which dunes are reducing in amplitude with increasing shear stress	1,000–2,400	1,400–4,000
Upper flow regime	Plane	1,500–3,100	——
	Standing waves	3,000–6,000	4,000–7,000
	Antidunes	5,000–42,000	6,000–15,000

is relatively small when the form of bed roughness consists of ripples and/or dunes. In the upper regime of flow the streams are capable of carrying much larger volumes of sediment per unit volume of water (see qualitative data in Table 7.1.3 suggested by Simons and Richardson [1963]). Some of the more useful concepts for estimating bed material discharge have been presented by Einstein [1950], Colby and Hembree [1955], Colby [1964], Simons *et al.* [1965], and Bishop *et al.* [1965]. For a more detailed treatment of the sediment problems encountered in designing and operating irrigation canals constructed in alluvium, refer to Simons and Miller [1966].

14. Summary

The basic concepts of fluid mechanics applicable to flow in open channels has been presented. For a more detailed treatment of hydraulics and fluid mechanics, see Albertson *et al.* [1960] and Albertson and Simons [1964] and other fluid-mechanics texts. Also, many valuable concepts pertinent to the design of hydraulic structures associated with the conveyance and distribution of water have been presented by the U.S. Bureau of Reclamation [1960, 1963, 1964].

REFERENCES

ALBERTSON, M. L., BARTON, J. R., and SIMONS, D. B. [1960], *Fluid Mechanics for Engineers;* (Prentice Hall, Englewood Cliffs, New Jersey), 568 p.

ALBERTSON, M. L. and SIMONS, D. B. [1964], Fluid Mechanics; In Chow, V. T., Editor, *Handbook of Applied Hydrology*, (McGraw-Hill, New York), Chapter 7, 49 p.

BAZIN, H. [1897], Etude d'une nouvelle formule pour calculer le debit des canaux decouverts; *Annales des Ponts et Chaussees*, Memoire No. 41, Vol. 14, Ser. 7, 4me trimestre, pp. 20–70.

BHOWMIK, N. G. [1965], *The Hydraulic Design of Large Concrete-lined Canals;* Thesis. Colorado State University, Fort Collins, Colorado.

BISHOP, A. A., SIMONS, D. B., and RICHARDSON, E. V. [1965], Total bed-material transport; *Proceedings of the American Society of Civil Engineers, Journal of the Hydraulics Division* 91 (HY2), 175–91.

COLBY, B. R. [1964], Discharge of sands and mean-velocity relationships in sandbed streams; *U.S. Geological Survey Professional Paper* 462-A, 47 p.

COLBY, B. R. and HEMBREE, C. H. [1955], Computations of total sediment discharge, Niobrara River near Cody, Nebraska; *U.S. Geological Survey Water Supply Paper* 1357, 187 p.

EINSTEIN, H. A. [1950], The bed-load function for sediment transportation in open channel flows; *U.S. Department of Agriculture Technical Bulletin* 1026; 71 p.

ELLIS, G. H. [1916], The flow of water in irrigation canals; *Transactions of the American Society of Civil Engineers*, Paper no. 1373, 1644–88.

GARBRECHT, G. [1961], Flow calculations for rivers and channels; Die Wasser-Wirtschaft, (Stuttgart), Parts I & II, 40–5 and 72–7. (U.S. Bureau of Reclamation Translation 402.)

HAUSHILD, W. L., SIMONS, D. B., and RICHARDSON, E. V. [1961], The significance of fall velocity and effective diameter of bed materials; *U.S. Geological Survey Professional Paper* 424-D, 17–20.

LACEY, G. [1958], Flow in alluvial channels with sandy mobile beds; *Proceedings of the Institution of Civil Engineers*, **9**, 145–64.

LANE, E. W. [1953], Progress report on studies on the design of stable channels by the Bureau of Reclamation; *Proceedings of the American Society of Civil Engineers*, **79**, 1–31.

LEOPOLD, L. B. and EMMETT, W. W., [1963], *Downstream pattern of River-bed scour and fill;* U.S. Geological Survey paper prepared for presentation of the Federal Interagency Sedimentation Conference, Jackson, Mississippi.

LEOPOLD, L. B. and MADDOCK, T. [1953], The hydraulic geometry of stream channels and some physiographic implications; *U.S. Geological Survey Professional Paper* 252, 56 p.

MANNING, R. [1889], On the flow of water in open channels and pipes; *Transactions of the Institution of Civil Engineering of Ireland*, **20**, 161–207. (Supplement, 1895, **25**, 179–207).

SIMONS, D. B. and ALBERTSON, M. L. [1963], Uniform water conveyance channels in alluvial material; *Transactions of the American Society of Civil Engineers*, **128**, 65–106.

SIMONS, D. B. and MILLER, C. R. [1966], Sediment discharge in irrigation canals; *Proceedings of the International Committee on Irrigation and Drainage, 6th Congress*, (New Delhi, India) Quest 20, Rep. 12, 20275–307.

SIMONS, D. B. and RICHARDSON, E. V. [1962], Resistance to flow in alluvial channels; *Transactions of the American Society of Civil Engineers*, **127**, 927–52.

SIMONS, D. B. and RICHARDSON, E. V. [1963], Forms of bed roughness in alluvial channels; *Transactions of the American Society of Civil Engineers*, **128**, 284–302.

SIMONS, D. B., RICHARDSON, E. V., and HAUSHILD, W. L. [1963], Some effects of fine sediment on flow phenomena. *U.S. Geological Survey Water Supply Paper*, 1498-G, 46 p.

SIMONS, D. B., RICHARDSON, E. V., and NORDIN, C. F. [1965], Bedload equation for ripples and dunes; *U.S. Geological Survey Professional Paper*, 462-H, 9 p.

U.S. BUREAU OF RECLAMATION [1960], *Design of Small Dams* (U.S. Government Printing Office, Washington, D.C.), 611 p.

U.S. BUREAU OF RECLAMATION [1960], *Design of Stable Channels with Tractive Forces and Competent Bottom Velocity;* Sedimentation Section, Hydrology Branch, Bureau of Reclamation, Denver Federal Center, Denver, Colorado.

U.S. BUREAU OF RECLAMATION [1963], *Hydraulic Design of Stilling Basins and Energy Dissipators* (Supplement of Documents, Washington, D.C.), Engineering Monographs **25**, 114 p.

U.S. BUREAU OF RECLAMATION [1964], *Design Standards No. 3: Canals and Related Structures;* Commissioner's Office. Denver Federal Center, Denver, Colorado.

7.II. Hydraulic Geometry

G. H. DURY

Department of Geography, University of Sydney

Hydraulic geometry is the graphical analysis of the hydraulic characteristics of a stream channel. These include width, depth, slope, discharge, velocity, bed material, and load. Discharge can be regarded as the only fully independent variable. Bed material is an independent variable in so far as its properties are determined by outcropping bedrock in the floor or sides of the channel; and it can also be regarded as an independent variable, since its properties are controlled by the geology of the catchment. Loose material, however, will be sorted during transit, and will also undergo reduction of size and change of shape. The remaining characteristics interact upon one another in a complex fashion. Whether they are to be classed as dependent or independent in a particular context may depend on the point of view. It will be convenient here to consider the whole series by groups, and to introduce at an early juncture the frequency-characteristics of discharge.

1. Interrelationships of discharge, velocity, depth, and width

Discharge is by definition the amount of water passing through the cross-section in unit time, expressed, for instance, in cubic feet per second. In what now follows depth is mean water depth and width is water surface width.

Consider an ordinary stream flowing in a single alluvial channel; and imagine to begin with a condition of low flow. Then, when discharge increases, so will depth and width increase. Part of the increase in depth will be accounted for by scour if the bed material can be shifted, although scour may not at once begin when stage rises (fig. 7.II.1). Part at least of the increase in width will result simply from the rise of water-surface level, but part at some sections will involve the temporary removal of bank material. Velocity also increases as discharge increases, in response to the enlargement of the cross-section and the associated reduction of average friction on the flowing water.

Variations in these and other characteristics at a given cross-section are called *at-a-station variations*. Although the foregoing verbal summary of their interrelationships is self-evident enough, there is more to the matter than this. Velocity, depth, and width can all be expressed as power functions of discharge (fig. 7.II.2). The prospect of bringing order into the variations of these four characteristics is, however, probably less important here than the indicated general inference, namely, that values of width, depth, and velocity cannot be

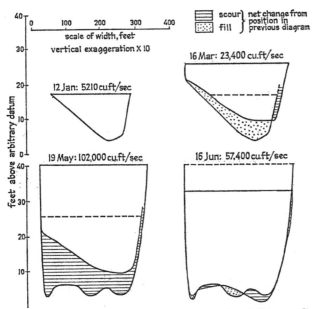

Fig. 7.11.1 Changes in size and shape of channel on the Colorado River at Grand Canyon, Arizona, during the passage of the December 1940–June 1941 flood (Freely adapted from Leopold and Maddock, 1953).

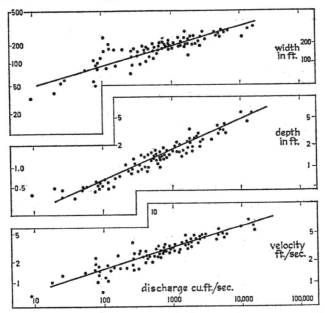

Fig. 7.11.2 Variation of width, depth, and velocity with varying mean annual discharge in the downstream direction: Powder River system, Wyoming and Montana (Adapted from Leopold and Maddock, 1953).

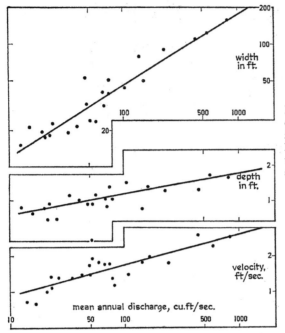

Fig. 7.11.3 Variation of width, depth, and velocity with varying discharge at-a-station: Powder River at Locate, Montana (Adapted from Leopold and Maddock, 1953).

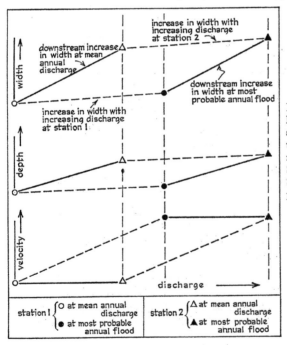

Fig. 7.11.4 Graphs of hydraulic geometry, showing at-a-station and downstream variations combined: diagrammatic, all scales logarithmic; Station 2 is downstream of Station 1 (Adapted from Leopold and Maddock, 1953).

used for comparative purposes unless they are in some way connected to the scale of discharge. It would be meaningless to compare one cross-section at low flow with another at high flow.

Variations of hydraulic characteristics along the length of a stream are called *variations in the downstream direction* or, more simply, *downstream variations*. Leopold and Maddock have investigated these, principally with reference to mean annual discharge, again finding that velocity, depth, and width can be expressed as power functions of discharge (fig. 7.II.2). However, the functional values usually differ from those obtained for at-a-station variations (fig. 7.II.3). It thus becomes possible to draw combined graphs (fig. 7.II.4) showing at-a-station variations in velocity, depth, and width with varying discharge, and downstream variations in velocity, depth, and width at discharge of given frequency.

Considerations of frequency need now to be outlined, in advance of further discussion of hydraulic geometry.

2. Magnitude-frequency analysis of discharge

Analysis of discharge-frequency necessarily involves magnitude, since very high discharges are rare (great magnitude correlates with low frequency) while low discharges are common (small magnitude correlates with high frequency). Several techniques of magnitude-frequency analysis are widely used: they include that developed by E. J. Gumbel, which will be outlined here.

Gumbel analysis depends on the statistical Theory of Extreme Values, one proposition of which is – crudely put – that records will eventually be broken. However severe an historical drought, there will sooner or later come a worse one; no matter how high a recorded flood, the future will bring one still higher. But although floods and droughts may come in spells, they do not come in cycles. In order to assess their time-relationship we need a non-cyclic method of analysis, such as that provided by Gumbel.

The operation of this method is best explained by an example. Table 7.II.1 lists in column 1 the highest peak discharge recorded in each year for a series of years at a single station. These peak discharges are called *floods*, whether or not they actually cause inundation. The series constituted by one peak per year is the *annual series*. The requirement is to define for each recorded flood the *recurrence-interval*, the average span of time between two successive floods of that particular magnitude. In column 2 of the table the floods have been rearranged in descending order of magnitude, and given ranking (serial) numbers. Recurrence-intervals are now calculated by means of the simple equation

$$r \cdot i \cdot = \frac{n+1}{r}$$

where $r \cdot i \cdot$ is recurrence-interval, n is the total number of items in the series, and r is the ranking order of a particular flood. The calculated recurrence-intervals for the series listed are given in column 3 of the table.

Observed magnitudes of floods are now plotted against recurrence-intervals on

TABLE 7.II.1 Annual floods on the Wabash River, at Lafayette, Indiana, 1924–57

Source: U.S. Dept. of Interior open-file Report

	1	2		3
Year	Peak discharge (ft³/sec)	Ranking Order	Discharge (ft³/sec)	Recurrence-interval (years)
1924	59,800	1	131,000	35
1925	63,300	2	93,500	17·5
1926	57,700	3	90,000	11·7
1927	64,000	4	74,600	8·8
1928	63,500	5	74,400	7·0
1929	38,000	6	73,300	5·8
1930	74,600	7	67,500	5·0
1931	13,100	8	64,000	4·4
1932	37,600	9	63,500	3·9
1933	67,500	10	63,300	3·5
1934	21,700	11	63,300	3·2
1935	37,000	12	62,000	2·9
1936	93,500	13	59,800	2·7
1937	58,500	14	58,500	2·5
1938	63,300	15	57,700	2·3
1939	74,400	16	52,600	2·2
1940	34,200	17	50,600	2·06
1941	14,600	18	46,600	1·95
1942	44,200	19	44,200	1·84
1943	131,000	20	41,900	1·75
1944	73,300	21	41,300	1·67
1945	46,600	22	41,200	1·59
1946	39,400	23	38,400	1·53
1947	41,200	24	38,000	1·46
1948	41,300	25	37,600	1·40
1949	62,000	26	37,000	1·34
1950	90,000	27	35,300	1·29
1951	50,600	28	35,000	1·25
1952	41,900	29	34,700	1·21
1953	35,000	30	30,000	1·17
1954	16,500	31	21,700	1·13
1955	35,300	32	16,500	1·09
1956	30,000	33	14,600	1·06
1957	52,600	34	13,100	1·03

special graph paper – Gumbel paper (fig. 7.II.5). If the data conform strictly to the Theory of Extreme Values, then the plotted points lie on a straight line. In practice, a certain amount of scatter s usual, and some plots give lines which are inflected either upwards or downwards. Nevertheless, they can still be used to read off magnitudes against recurrence-intervals, and vice versa. Against the recurrence-interval of 50 years is read off the magnitude of the 50-year flood, the peak discharge expectable as an annual maximum once in a 50-year span. Emphatically, this does *not* mean at regular intervals of 50 years. There is nothing to prevent the occurrence of two 50-year floods in two successive years. Furthermore, some 50-year spans must inevitably include 100-year floods (on the average, one such span in two) 500-year floods (one in ten), and 1,000-year floods (one in twenty).

For preference, the period of record should be twice as long as the greatest recurrence-interval for which flood magnitude is obtained. Few series of records are as long as 100 years, but for many stations the magnitudes of the 25-year,

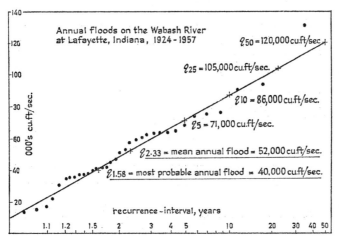

Fig. 7.11.5 Example of a Gumbel graph: scales simplified and graph face omitted (cf. Table 7.11.1).

10-year, and 5-year floods can be satisfactorily defined. These floods come within the practicable limits of much engineering work, flood control, and flood alleviation. For present purposes, lesser recurrence-intervals and lesser floods are of main interest. The 2·33-year flood is *the mean annual flood*; the 2-year flood is the *median annual flood*, equalled or exceeded in one year in two; and the 1·58-year flood is the *most probable annual flood*. These are symbolized respectively as $q_{2·33}$, $q_{2·0}$, and $q_{1·58}$.

Both below and in a later section it will be necessary to discuss discharge in relation to stage – specifically, in relation to discharge at the bankfull stage. Bankfull discharge is symbolized as q_{bf}. If it were possible it would be useful to locate bankfull discharge on the magnitude-frequency scale, but complications occurring on natural rivers, quite apart from those due to engineering works, make the desired act of location very difficult. But the matter can be approached in another way. Peak floods below the median value disappear from half the record of annual peaks, since they are by definition equalled or exceeded in one year in two. Similarly, peaks equivalent to the most probable annual flood vanish from more than half of the annual series. Now when account is taken of flood peaks smaller than the annual maxima it can be shown that, in addition to being the most probable annual maximum, the discharge equivalent to $q_{1.58}$ is likely to occur once every year. These seem good grounds for suggesting that $q_{1.58}$ = natural bankfull discharge. It is possible to take the matter further, saying that natural bankfull discharge = channel-forming discharge, on the grounds that analysis of such matters as sediment delivery shows that channels are shaped not by events of great magnitude and low frequency but by those of modest magnitude and high frequency. In addition, it is at bankfull stage that water is in contact with, and is acting on, the whole perimeter of the channel.

3. Interrelationships of slope, discharge, and velocity

Downstream channel slope on most rivers tends to decrease from source to mouth, giving longitudinal profiles a concave-upward form. The basic reason for a downstream decrease of slope is a downstream increase in channel efficiency.

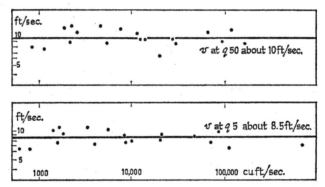

Fig. 7.II.6 Velocity-relationships on the Yellowstone–Missouri–Mississippi system, at the five-year and fifty-year floods (Adapted from Leopold, 1953).

This increase is a response to the increase in channel size which results from the entry of tributaries. The combined effects of relief distribution, basin shape and size in relation to the layout of the stream network, and runoff characteristics frequently ensure that the concave-upward profile tends to be approximately logarithmic. Alternatively, the logarithmic form can be regarded as the outcome of statistical probability.

Leopold began the study of velocity in relation to slope with an analysis of rivers in the U.S. Mid-west, for which he used mean discharge as a basis of reference. This discharge corresponds to below-bank stages. Leopold showed that, at mean discharge, velocity through the cross-section can actually *increase* downstream. Later work on the Yellowstone–Mississippi revealed that there is no tendency for velocity to change along the length of this system, despite a downstream decrease in slope from 100 to 0·5 ft/mile, either at the 50-year or at the 5-year flood (fig. 7.II.6).

It appears to follow that constant velocity along the length of the stream is attained somewhere between the below-bank stage of mean discharge and the modest overbank stage of the 5-year flood. Theoretical analysis of stream behaviour at bankfull combines with pilot empirical studies of velocity at $q_{1.58}$ to suggest that constant downstream velocity is attained at bankfull or at the most probable annual flood, according to which is in question. That is to say, channel slope tends to be adapted to promote constant velocity downstream at channel-forming discharge.

These observations and this conclusion make nonsense of the assertion – unfortunately still a common one – that streams rush hurriedly down their mountain courses but move sluggishly across the plains. This could be so only if there were no downstream change of channel size.

4. Interrelationships of slope, width, depth, and bed material

Many longitudinal profiles are far from regular: they include reaches where slope increases downstream. Profiles consisting of segments of concave-up curves may have been developed under the control of intermittent falls in the level of the sea relative to the land. But sea-level change is by no means the only possible cause of downstream steepening of gradients and of segmented profiles. Any factor which reduces channel efficiency may be compensated for by some equal but opposite factor or group of factors. The readiest mode of compensation is often an increase in slope.

Large channels are, as stated, more efficient than small channels. If therefore a single channel subdivides in the downstream direction into two or more lesser channels the subdivision may well be accompanied by an increase of slope. The most efficient channel shape is semicircular. Since alluvial channels are usually shallower than semicircular, anything which causes them to become wider and shallower will reduce their efficiency and tend to promote a compensating increase in slope. Schumm has investigated the influence of particle-size in the materials of the beds and banks of alluvial channels, finding that a high proportion in the silt and clay grades produces a low width/depth ratio, whereas a low proportion of silt-clay is associated with great width in relation to depth. The two kinds of shape are, respectively, efficient and inefficient.

At the other end of the range of possible shapes are channels which are mere slots in bedrock, excavated vertically down belts of weakness but confined by strong vertical walls. These, too, are markedly inefficient, and likely to appear on longitudinal profiles as reaches where slope increases downstream.

Whether in alluvium or in bedrock, channels range considerably in their frictional drag on the flowing water. It is partly for this reason that the well-known equation of Manning, which relates velocity to size, shape, and slope, also includes a roughness factor. Roughness in this context includes not only physical obstruction such as that produced by irregularly jutting masses of rock in place but also the effect of particle-size in the beds and banks. To this extent it overlaps with the properties investigated by Schumm. In addition, roughness includes the frictional drag produced by material in transit. Once again, the greater the roughness, the greater the slope required to promote a particular velocity at a given discharge.

It follows from all this that any downstream increases of slope which may be detected by survey or from accurate maps cannot be fully understood unless their possible causes have been checked in the field.

5. Grade

In its application to rivers, the concept of grade has often proved elusive and sometimes confusing. It calls for a balance between the capacity to do work (in particular, to transport load) and the amount of work to be done – e.g., the quantity of load supplied. A graded profile is supposed to be produced when capacity and performance are matched. However, the whole of the very bulky early debate on grade is rendered largely meaningless by lack of reference to magnitude-frequency of discharge. Use of the concepts of magnitude-frequency, and especially of channel-forming discharge, directs attention not so much to grade as to steady-state conditions. In any event, it has been pointed out that slope is determined with reference to a whole series of additional variables, so that a graded profile, however defined, need not be a smooth continuous curve throughout.

6. Load in relation to material of the bed and banks

It is useful for a number of purposes to separate material in transit into somewhat arbitrary fractions, according to its mode of progression along the channel. *Bedload* consists of material too coarse to be supported in the flowing water for any appreciable time. It moves by sliding, rolling, or saltating, continually coming to rest and thus reverting to bed material. *Suspension load* is carried in the body of the current. The degree of coarseness of particles varies with velocity, and especially with the degree of turbulence. Streams at bankfull can be assumed to have turbulent flow, with rising and descending eddies. At a velocity of 3 ft/sec, which is modest for flow at bankfull stage, transport of sand is approximately uniform through all parts of the cross-section, and a stream can deliver 10 tons per day for every foot of filled channel width at a given range of depth.

Some writers separate off as *wash load* that part of the suspended load which, although consisting of solid particles, is so fine-grained that its settling velocity is very small or nil. This portion is carried in colloidal suspension. *Solution load* is by definition transported in the dissolved state.

Most of the available information on transport of load by rivers deals either with suspended load or with solution load. Movement of bedload is remarkably difficult to measure in field conditions. On the other hand, a great deal is known about the tendency of particles larger than sand size to become reduced in dimensions with progression downstream, and also about the critical stress needed to move particles of a given size. It is fully clear that bedload movement tends to increase as velocity increases: that is – up to the bankfull stage at least – as discharge increases. Transport in solution varies widely in its effectiveness. It is controlled at least in part by the geology of catchments and by climate. In some arid (salty) and humid-tropical environments it can exceed transport in suspension, just as it can in catchments where the rocks are markedly soluble. The relationships of solution-load to discharge are more complicated than those of bedload, for solution-load concentration is likely to be low at times of high discharge, whereas total transport in solution is similarly low at times of low discharge. Generally speaking, the greatest effectiveness of transport in solution seems likely to be attained at stages intermediate between low and very high.

Suspended load has been well studied. At many stations on many rivers it

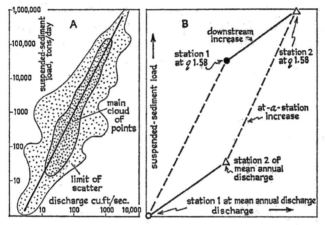

Fig. 7.11.7 (A) Variation of suspended-sediment load with varying discharge, Powder River at Arvada, Wyoming (Adapted from Leopold and Maddock); (B) Combined at-a-station and downstream variations of suspended-sediment load with varying discharge; diagrammatic, all scales logarithmic: Station 2 is downstream of Station 1 (Compare Fig. 7.11.2) (Adapted from Leopold and Maddock, 1953).

increases as a power function of discharge, increasing as a rule far more rapidly than discharge does (fig. 7.11.7(a)). As for width, depth, and velocity, so for suspended-sediment load, a combined graph can be drawn to show at-a-station and downstream variations (fig. 7.11.7(b)). Like the concentration of solution-load, the concentration of suspended-sediment load is apt to be less at times of very high flow than at times of moderate flow, simply because the suspended material is disseminated through a great volume of water. Magnitude-frequency analysis of sediment transport shows that the greatest total transport is effected by discharges of modest magnitude and high frequency, similar to those recognized as responsible for shaping the channels themselves.

REFERENCES

LEOPOLD, L. B. [1953], Downstream change of velocity in rivers; *American Journal of Science*, 251,606–624.

LEOPOLD, L. B. and MADDOCK, T. JR. [1953], The hydraulic geometry of stream channels and some physiographic implications; *U.S. Geological Survey Professional Paper* 252, 57 p. (The basic reference on hydraulic geometry.)

LEOPOLD, L. B., WOLMAN, M. G. and MILLER, J. P. [1964], *Fluvial Processes in Geomorphology* (W. H. Freeman and Co., San Francisco and London), 522 p. (This incorporates much relevant work, especially in Chapters 6 and 7.)

WOLMAN, M. G. [1955], The natural channel of Brandywine Creek, Pennsylvania; *U.S. Geological Survey Professional Paper* 271, 56 p.

WOLMAN, M. G. and MILLER, J. P. [1960], Magnitude and frequency of forces in geomorphic processes; *Journal of Geology*, **68**, 54–74.

8.II(i). The Geomorphology and Morphometry of Glacial and Nival Areas

IAN S. EVANS

Department of Geography, Durham University

Glacial and nival hydrology is particularly sensitive to temperature and radiation, and hence latitude, altitude, and aspect are especially important influences on glacial and nival landforms. The latitudinal sequence polar ice-cap, polar desert, tundra, boreal forest is matched by the altitudinal zonation nival, subnival, alpine, sub-alpine, although high altitudes differ from high latitudes in the relationship of seasonal and diurnal fluctuations and in radiation, wind, and weather. In addition, slope direction and relative position on a slope are very important in mountain areas, and complicate the pattern of zonation. For example, whereas small glaciers are zonal forms found at the snowline for their aspect, large glaciers are azonal, and a large actively fed glacier in a steep valley may transgress down into a much warmer zone. Slope gradient and aspect affect the surface receipt of precipitation and radiation. This leads to asymmetry of nival balance (affecting intensity of glaciation) and of temperature (affecting freeze–thaw cycles, chemical weathering, and availability of meltwater). The greatest direct radiation at a given angle of latitude is received by an equatorial-facing slope of similar angle, and the least by steep north-facing slopes. This effect is most marked in middle latitudes and in clear conditions. Since west-facing slopes are heated in the afternoon when the air temperature is highest, south-west-facing slopes tend to be warmest and north-east-facing slopes coolest. This probably explains the general mid-latitude tendency for glaciers and cirques to face preferentially north-eastward. Wind drifting from smooth summits and plateau areas produces snow accumulation on lee slopes, and in the west-wind belt this encourages east-facing glaciers and cirques. However, since falling precipitation travels with the wind, its incidence is greater on a surface perpendicular to its direction of travel (i.e. of westerly aspect). In rugged terrain this effect roughly counterbalances wind drifting, leaving the radiation factor dominant.

1. Cirques

A cirque is defined morphologically as a large steep-sided hollow, open on the downstream side but essentially closed upstream by a steep, arcuate headwall below a divide; its floor slopes more gently than its sides, and also more gently

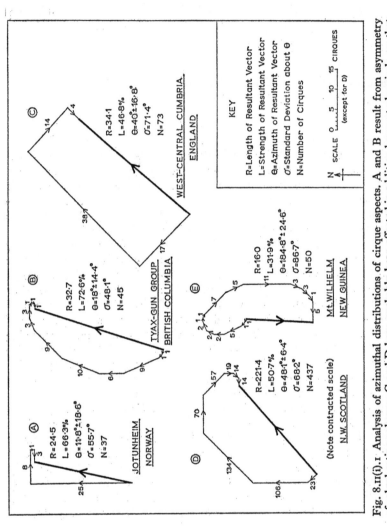

Fig. 8.11(i).1 Analysis of azimuthal distributions of cirque aspects. A and B result from asymmetry of solar heating, whereas C and D have probably been affected in addition by westerly winds, so that they have weaker azimuthal concentrations. E is a still more dispersed distribution from an equatorial region where solar heating is more uniform.

than the succeeding slope downstream. Little ice flows, or flowed, in from outside the cirque. The convex change in slope at the downstream end of a cirque floor is known as a threshold (or sill). Especially if it is ice-moulded, this feature helps to distinguish glacial cirques from comparable forms developed in nonglacial areas of favourable structure, where mass movement broadens valley heads faster than stream cutting deepens them. Also a cirque floor does not normally coincide closely with a stratigraphic surface.

The influence of aspect is greatest in areas of 'marginal glaciation', where the regional snowline is not far below mountain crests. Its effect has remained important even where the glaciation intensified subsequently. Glaciers facing north-east aided the development of cirques by rapidly removing frost-wedged debris and simplifying accumulation basins. This extended north-east-facing valleys by pushing back divides until they were located far to the south-west of range centre-lines. If later lowering of the snowline permitted glacier development on south-west-facing slopes as well the same opportunities for catchment accumulation and cirque development were often not available there, the divide could not be pushed back, and the majority of cirques continue to face north-east. Figure 8.II(i).1 shows cirque-aspect statistics for several groups of cirques influenced chiefly by radiation, and for others which are influenced in addition by wind. Even within cirques, north-east-facing slopes tend to be steeper, so that cirques facing other azimuths may be internally asymmetric. Azimuthal distributions of gradient are complicated by the development within cirques of gently sloping floors as well as steep walls. A range in altitude which includes both shows greater gradient variance at the preferred azimuth of cirques. Mean gradient is affected when the altitudinal range includes mostly walls or mostly floors. The altitude of cirque floors is an altitude of minimum gradient and maximum surface area, especially for preferred cirque azimuths.

In plateau areas cirques are often isolated and easy to delimit. But in more thoroughly dissected terrain their coalescence creates difficult problems. If several hollows, separated by distinct spurs, coalesce to share the same threshold they may be considered individual cirques if the floor can be partitioned easily between them (fig. 8.II(i).2(a)). If, however, the spurs are much smaller, minor protrusions from a continuous backwall, the feature is a single cirque (fig. 8.II(i).2(c)). A distinction between 'valley-side' and 'valley-head' cirques is not useful, since all cirques inevitably concentrate drainage and are necessarily the heads of small valleys. A more meaningful dichotomy is between 'closed' cirques (*cirques en fauteuil*), whose floors are essentially basins cut in rock, and 'open' cirques (*cirques en van*), whose floors slope generally outwards, but at an angle less than that of the slopes below and above. Open cirques are more common well above the snowline, in major glacial source areas of high relief; whereas closed cirques are more characteristic of areas of marginal glaciation, and their floor altitudes have been shown to bear an apparently reasonable relationship to the regional snowline. Alternatively, on weak rocks such as shales cirques are commonly open; whereas on massively jointed igneous rocks, limestones, and sandstones they more often have rock basins or at least flat floors.

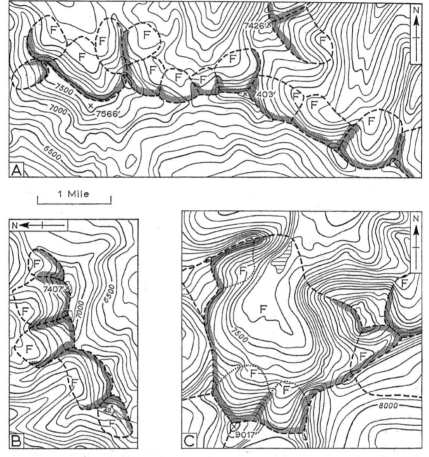

Fig. 8.11(i).2 Cirque definition and variation, in the northern part of the Bridge River District, south-western British Columbia (Heights in feet). Delimitations are based on ground and air photographs (Maps reproduced by permission of the British Columbia Department of Lands, Forests and Water Resources).

A. Coalescent individual cirques at the sources of Eldorado and Taylor Creeks, Tyax–Gun Group. Cirque floors (F) are adjacent but distinct. Contours are shown as continuous lines, cirque outlines as dashed lines, and the upslope margins of the cirques are shaded.

B. Shallow cirques south-east of Eldorado Creek, showing (*bottom*) the definition of cirque plan closure (i.e. the range in azimuth of the longest contour: 42°).

C. A large diversified cirque, deep both in plan and in profile, at the source of the southern fork of Blue Creek, Shulaps Range. The plan closure is 271°. There are several tributary cirques, each with weakly developed floors and thresholds, now almost obscured by detritus.

It seems that strong rock is necessary to sustain a threshold with a reversed slope.

Cirque geometry can be described quantitatively, firstly, with reference to closure in long profile (i.e. in vertical section) as the difference between the maximum backwall slope and the minimum outward floor slope (which may be negative, i.e. reversed) (fig. 8.11(i).3); and secondly, the closure in plan may be defined as the range in azimuth of the longest contour within the cirque. A 'plan closure' less than 90° indicates a cirque which is only a shallow indentation in the mountainside (fig. 8.11(i).2(*b*)), while one in excess of 180° denotes a cirque closed

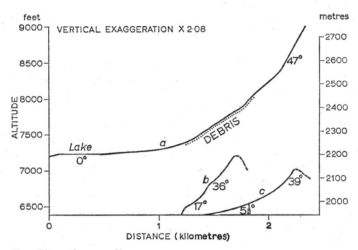

Fig. 8.11(i).3 Cirque long profiles.

A. Large cirque shown in Fig. 8.11(i).2(*c*). This has quite a large profile closure which is at least 47° (i.e. 47°−0°).

B. The poorly developed westernmost cirque shown in Fig. 8.11(i).2(*b*). This has the relatively small profile closure of 19° (i.e. 36°−17°).

C. A simple north-east-facing cirque in the Tyax–Gun Group. Profile closure is $33\frac{1}{2}°$ (i.e. 39°−5$\frac{1}{2}$°).

on more than three sides (fig. 8.11(i).2(*c*)). These two parameters are the most important measures of cirque shape. Dealing with a single cirque long profile is troublesome, firstly, in the need to define a cirque 'centre-line', and secondly, in that the profile selected may not be characteristic of the cirque as a whole. At the loss of some detail, this may be avoided by constructing a clinographic curve showing the altitudinal variation of mean gradient within the cirque. Profile closure can be expressed as the difference between maximum and minimum altitudinal mean gradient.

The measurement of cirque dimensions is more difficult, since width, length, and especially depth are sensitive to the influence of external topography. Cirque area is one simple expression of cirque size; the length of the longest unbroken contour within a cirque is another, increasing with elongation but less affected by cirque outline. There is no obvious upper limit to cirque size, except that cirque coalescence takes time, and pre-glacial topography rarely favoured early formation of large cirques; cirque growth in area is necessarily slower when opposed by other growing cirques on several sides. While cirque size may, of course, vary with length of glaciation (largely controlled by altitude relative to snow lines), the passage of time does not have the cardinal importance often assumed. Tectonic environment and pre-glacial topography account for most of the differences in 'stage of glacial dissection'. Hobbs's view that cirques normally subdivide and become more intricate as they develop is now generally rejected in favour of the view that spur elimination leads to simplification of cirque form. Irregularities may be accentuated if they relate to rock structure, or if the cirque glacier subdivides so that indentations in the cirque wall continue to be sapped while spurs do not.

While cirque rock basins are produced solely by glacial erosion, the development of the surrounding steep walls is more complex. The process observed on them today is rockfall following loosening by the expansion of water freezing in cracks. The fallen rocks accumulate below and will eventually bury the whole cliff, which developed when a glacier carried away such fragments and permitted the attack by freeze–thaw to continue. A more controversial possibility is that the glacier actually undercut the cliff, as might the sea. Freeze–thaw action might have been accelerated at the glacier margin by an increased availability of water in the bergschrund or rimaye. Temperature oscillations there would be reduced, but when abundant water is available a single cycle of deep freeze and thaw is probably more effective than many rapid oscillations. Meltwater may sometimes penetrate under the ice and cause freeze–thaw at considerable depths, followed by 'plucking', the incorporation of loosened blocks into the moving ice.

2. Nivation

It has been suggested that cirques develop from nivation hollows. Although nivation is important early in the transformation of a steep gully or landslide scar into a cirque, most nivation hollows are on a much smaller scale, and lack the initial relief to accumulate sufficient snow to develop into cirques. They are generally found on gentle slopes, especially where a waste mantle has accumulated, or on weak rocks, where they are commonly elongated parallel to rock structure. Though not proven, their causal relation to snowbanks is reasonably inferred from their spatial association with long-lasting snowpatches, where the concentrated availability of meltwater favours frost disintegration of the waste mantle, as well as solifluction, earthflows, and surface wash on bare areas uncovered late in summer. Eluviation of fine material through the waste mantle permits the development of closed depressions.

Avalanches are a quite different manifestation of nival action. Their importance in areas of high relief and snowfall has not been adequately recognized. When a snow cornice breaks and sweeps down a cliff it cleans off a considerable amount of rock already loosened by freeze–thaw. Such slab avalanches are most common on lee slopes, and permit them to remain steep as they retreat. But slush avalanches are also very effective erosive agents, even on much gentler slopes. Hence, even windward or sunny slopes can be indented by dendritic networks of shallow gullies ('chutes'). These are also followed by streams and rockfalls, but avalanches are probably the major factor in their genesis.

3. Glacial erosion

Erosion by valley glaciers depends upon rock erodibility, the amount of basal debris, the 'hydraulic geometry' of the glacier channel, and basal ice conditions, such as velocity, viscosity, and temperature. These relationships are very complex, and an understanding of how glaciers erode is a long way off. Landforms related to past glaciation suggest that erosion increased where glaciers became either thicker or faster. Fast-flowing glaciers can achieve considerable abrasion, especially if they slide over their bed; however, it is plucking which seems to account for most glacial erosion. This is greatly facilitated if glacier ice can freeze on to a loosened block: refreezing has been observed at shallow depths, but is more difficult to envisage at depth where pressure is more even and, in a temperate glacier, ice is kept at pressure melting point. Changes in pressure through time are important in causing oscillations between 'cold' and 'temperate' ice. Sections of frozen bed material may thus be attached to the glacier sole and removed from their setting at a time of basal melt (i.e. increased pressure). This may explain the lifting of very large blocks of chalk, found in the till of Norfolk and Denmark. In other cases the push of the ice and especially of basal moraine may pry out loosened fragments. On a large scale, the $4\frac{1}{2}$-km-long limestone island of Osmussar in north-west Estonia seems to have been turned through 16°, over a shale base.

Glacial erosion of incoherent or jointed rock is more comprehensible than the continued erosion of hard, massive rocks to great depths. The development or opening of joints parallel to the surface is favoured by expansion due to unloading (i.e. the removal, by erosion, of the weight of superincumbent rock). It has recently been found that horizontal stresses in the earth's crust are several times greater than vertical loads. Hence when a deep trough is cut its walls expand laterally into it, releasing compression in rocks at this level, but increasing it in lower layers below the trough. This excessive stress concentration probably causes up-arching into the trough, fracturing the rock so that it can be removed by the glacier. As the trough is deepened the stress concentration becomes even greater, counterbalancing the weight of an increased thickness of ice. In this process we may at last have an explanation of why glaciers can cut deep rock basins in hard rock, and why erosion increases with ice thickness so long as ice is confined in a channel.

It is often suggested, even now, that glaciers merely transport material already loosened and debris fallen from the slopes above. While these two sources are important, they completely fail to account for the excavation of deep rock basins, which are the most distinctive result of glacial sculpture. Such basins and troughs cannot be due to glacial removal of rock deeply rotted in the Cenozoic, since their distribution relates to the former glacier system and their walls are generally of sound rock; their pattern is quite different from that of lows in a 'weathering front'. Furthermore, the great quantity of 'rock flour' carried by streams draining from glaciers demonstrates how much abrasion and corrasion is proceeding beneath even small glaciers. The scepticism of some workers relates to the small quantity of moraine currently found in some glaciers. It has been estimated that the Mer de Glace removes only a 'few dozen' cubic metres/year, which lowers its basin by a few microns; that the Glacier de Saint-Sorlin (Grandes Rousses), on weaker rock, lowers its basin 0·1 mm/year; and that there is a lowering of 0·05 mm/year in East Antarctica, which is more rapid than fluvial denudation in areas of low relief, but much less than that in semi-arid or mountain areas. On the other hand, the glacial rock flour carried by streams draining Muir Glacier, Alaska, corresponds to an average loss of 19 mm/year from the area beneath the glacier. Apart from the great approximations involved in such estimates, it is unreasonable to expect modern glaciers, however large, to erode as much as more extensive past glaciers. To the extent that valleys currently glaciated have been modified to suit larger past glaciers, little contemporary modification is either likely or necessary. Thus the most recent Yosemite Glacier in California cut only 220 m into loose sediments filling the 550-m-deep closed basin which was cut into granite by its larger predecessor. Similarly, several Alpine glaciers are floored by a layer of old moraine or proglacial deposits often as thick as the overlying ice. The floor of the present glacier channel reflects the bedrock surface, since both have been cut by similar glacier systems.

The reality of glacial erosion is confirmed most strikingly by the way that channels of former glaciers, as reconstructed on the basis of their deposits, striations, and ice-moulded forms, are well adjusted to their estimated ice discharge. As in fluvial hydraulic geometry, the effect of an increase in discharge (mean velocity times cross-sectional area) is shared out among velocity, width, and depth, all of which tend to increase. Where two glaciers join, their surface must slope continuously downstream (if flow is to continue), and if the bed inherited from fluvial action slopes gently, glacier depth cannot increase downstream; hence velocity increases excessively, increasing erosion until an equilibrium depth is reached, with an increased bed slope at the confluence. This effect is most marked near the source of a glacier, where ice from several cirques joins to form a trunk glacier, leading to an abrupt increase in discharge and a steep drop to a 'trough-head'. A closed rock basin is more likely to form in the ablation zone, where ice moves towards the glacier surface to compensate for the the net surface loss by ablation. Where the surface slopes gently, or near the snout, there may actually be a vertically upward flow component. The downstream reduction in discharge may then be accommodated by a rise in the glacier floor. Hence the largest rock basins are often just above glacier termini, and that of Sogne Fjord (Norway) is at least 1,150 m deeper than its terminal threshold. The rotational flow of cirque glaciers is a special case of this phenomenon when a cirque basin is cut beneath the ablation zone, where ice-flow direction changes from downward to upward. Similarly, where a valley broadens, especially at the margin of a mountain range, the sudden increase in glacier width produces a decrease in depth and velocity, encouraging the formation of a reversed slope closing a rock basin.

Other closed rock basins, so characteristic of glaciated regions, are due to differential rock erodibility; in particular, basins are cut where joints are more closely spaced. Some basins coincide with shear belts of shattered rock, others with weak dykes. Indeed, the locations of basins formed under an ice-sheet are controlled principally by rock structure.

Where a small glacier joins a large one, the result is a negligible step beneath the trunk glacier but an abrupt step beneath the small one (which becomes inset). This is left as a 'hanging valley', but in reality it is a 'hanging paleochannel', for the surfaces of the two glaciers were originally accordant. In many ways glacial hydraulic geometry is similar to fluvial hydraulic geometry, except for the larger cross-section of the channels due to the fact that ice is more viscous than water. Thus the parabolic shape of a glaciated valley (i.e. paleochannel) compares with the cross-section of a river channel, but glacier channels show less variation from this mean form because their laminar flow tends to reduce, rather than exaggerate, channel asymmetry. Laminar flow, with a lack of cross-currents, also leads to the smoothing of sharp pre-glacial valley curves, so that glacier troughs are either straight or curving broadly. There is no evidence that glaciers meander.

As glaciers grow, they cease to be confined in the previous valley system. Ice streams overflow low points in divides to escape to valleys where the ice surface is lower. Rapid flow wears down the divides, creating 'through valleys'; and commonly a rock basin is cut across a former divide, where ice flow was more constricted and therefore needed a deeper channel. Greater ice surface slope on the down-stream side permits greater erosion than where the ground slopes upstream; hence the low point which becomes the post-glacial divide is displaced 'up-ice', often a considerable distance. Glacial 'transfluence' of divides becomes more widespread as ice overwhelms the landscape, and a mountain ice-sheet produces radiating valleys by deepening suitably oriented valley sections and integrating them by divide breaching. In areas of low relief glacial erosion is less concentrated and an ice-sheet does not carve new valleys in geologically homogeneous plains, except that deep basins may result where flow is locally concentrated by a topographic obstruction.

Deglaciation produces rapid landscape modification as fluvial and slope processes begin readjusting the glaciated landscape to a new fluvial equilibrium. Unstable, newly exposed rock walls collapse, and gorges (usually initiated by sub-glacial streams) are cut into oversteepened slopes. Renewed glaciation encounters a changed situation and makes further changes towards its own equilibrium, so that erosion by repeated glaciation may well exceed that by a single long glaciation.

4. Streamlined and depositional forms

Ice tends to mould large bed obstacles into streamlined forms, by erosion and deposition combined in various proportions. The most distinctive form is the drumlin, of the order of a kilometre long and less than half as broad; in plan rounded upstream and pointed downstream; and composed commonly of till, but sometimes of stratified drift or even bedrock. The long axis is aligned in the direction of ice flow, and the upstream end is steeper. Drumlins are probably formed principally by erosion of older glacial deposits, and their elongation may depend on a force/resistance ratio (i.e. it increases with basal ice velocity and decreases with till resistance). Drumlins are usually found in large groups (within which elongation tends to be consistent) not far upstream from terminal moraines, and were formed especially beneath glacier lobes with divergent flow. Though similar forms ('crag and tail') are produced by till deposition in the lee of a bedrock obstacle, drumlins are not necessarily related to such obstacles, but rather to a regular instability at the ice/till interface.

In the ablation zone, and especially during glacier wastage, meltwater is a powerful geomorphic agent. At the glacier bed it moves under considerable and rapidly changing pressure, and is a very effective erosive agent. Sudden discharges caused by the bursting of subglacial barriers, for example, as water in ice-dammed lakes reaches a critical level and drains out under the ice, are particularly erosive. Major subglacial meltwater channels may be distinguished from subaerial ones in having 'up-and-down' long profiles and being discordant with local topography, as when they are cut through spurs. On the glacier surface, meltwater washes fine-grained material out of supraglacial moraine, which is deposited as loose 'ablation till', distinct from the compact 'lodgement till' deposited subglacially under pressure and with fine material retained. Meltwater deposits some material as coarse foreset beds in subglacial channels, forming linear gravel ridges (eskers) when the ice walls melt away, but much is taken farther to form great depositional plains of outwash alongside and downstream from the glacier, mixing with an increasing amount of non-glacial material. Near the glacier, outwash is not only coarser and less well sorted but also disturbed by loss of support when the ice melts, so that it can be distinguished as 'ice-contact stratified drift'. During glacier wastage outwash may cover much of the near-stagnant glacier tongue, accumulating in meltwater channels, crevasses, and other hollows in the ice. Ice melting produces inversion of relief: 'kames' are the positive forms once surrounded by ice, while 'kettles' are the negative ones, where a block of ice was surrounded by drift deposits. Lateral accumulations between a glacier and a valley-side become 'kame terraces'. Where the transition from active to near-stagnant ice is rapid, thrust-planes often carry debris to the surface, where differential ablation produces linear ridges. From these ridges, debris slumps down over the glacier snout and on to the proglacial outwash plain as sheets of 'flow till'. In this way, till and outwash can be interbedded without any readvance of the ice margin. On valley glaciers much till is deposited by ablation at the ice margin, building an end moraine. The inside of this is a cast of the glacier margin, the outside often a talus at the angle of rest. Such a moraine may rim the glacier as far upstream as the firn line, but it is better developed where ablation is greater; it is often lacking on the shady side of the valley. The lateral parts of the moraine are often better preserved, since meltwater streams breach the terminal moraine at the glacier snout: complete arcuate ridges are typical of the smallest glaciers, which produce less meltwater. In lowlands the till component is often small, and the moraine is a series of deltas, the head of an outwash plain, or a bulldozed ridge of various deposits. End moraines are formed by the culminations of readvances, rather than pauses in retreat. Once formed, their influence may keep the glacier margin stationary for a longer period, by containing minor oscillations.

REFERENCES

AHLMANN, H. W. [1919], Geomorphological studies in Norway; Geografiska Annaler, 1, 1–148 and 193–252.

ANDREWS, J. T. [1963], Cross-valley moraines of north-central Baffin Island: a quantitative analysis; Geographical Bulletin, 20, 82–129.

BLACHE, J. [1952], La sculpture glaciaire; Revue de Géographie alpine, 40, 31–123.

BOULTON, G. S. [1967], The development of a complex supraglacial moraine at the margin of Sørbreen, Ny Friesland, Vestspitsbergen; Journal of Glaciology, 6 (47), 717–35.

CHARLESWORTH, J. K. [1957], The Quaternary Era with special reference to its glaciation (London), 1,700 p.

CHORLEY, R. J. [1959], The shape of drumlins; Journal of Glaciology, 3 (25), 339–44.

CLAYTON, K. M. [1965], Glacial erosion in the Finger Lakes region (New York State, U.S.A.); Zeitschrift für Geomorphologie, n.f. 9, 50–62.

COTTON, C. A. [1942], Climatic Accidents in Landscape-Making (Christchurch, New Zealand), 354 p.

DAHL, R. A. [1965], Plastically sculptured detail forms on rock surfaces in northern Nordland, Norway; Geografiska Annaler, 47A, 83–140.

DAVIS, N. F. G. and MATHEWS, W. H. [1944], Four phases of glaciation, with illustrations from south-western British Columbia; Journal of Geology, 52, 403–13.

EMBLETON, C. and KING, C. A. M. [1968], Glacial and Periglacial Geomorphology (London), 608 p.

EVANS, I. S. [In preparation], Measurement and Interpretation of Asymmetry in Glaciated Mountains.

FLINT, R. F. [1957], Glacial and Pleistocene Geology (New York), 553 p.

GEIGER, R. [1965], The Climate Near the Ground (Cambridge, Mass.), 611 p.

GILBERT, G. K. [1904], Systematic asymmetry of crest lines in the High Sierra of California; Journal of Geology, 12, 579–88.

LEWIS, W. V., Editor [1960], Norwegian Cirque Glaciers; Royal Geographical Society Research Series, No. 4, 104 p.

LINTON, D. L. [1963], The forms of glacial erosion; Transactions of the Institute of British Geographers, 33, 1–28.

LLIBOUTRY, L. [1965], Traité de Glaciologie, t.2. Glaciers-variations du climat-sols gelés (Paris), pp. 429–1040.

[168] INTRODUCTION TO FLUVIAL PROCESSES

MATTHES, F. E. [1900], Glacial sculpture of the Bighorn Mountains, Wyoming; *United States Geological Survey, 21st Annual Report*, Part 2, 167–90.

MATTHES, F. E. [1930], Geologic history of the Yosemite Valley; *United States Geological Survey Professional Paper* 160, 137 p.

RAPP, A. [1960], Recent development of mountain slopes in Kärkevagge and surroundings, North Scandinavia; *Geografiska Annaler*, 42, 65–200.

SEDDON, B. [1957], Late-glacial cwm glaciers in Wales; *Journal of Glaciology*, 3 (22), 94–9.

SISSONS, J. B. [1967], *The Evolution of Scotland's Scenery* (Edinburgh), 259 p.

TRICART, J. and CAILLEUX, A. [1962], *Le modelé glaciaire et nival* (Paris), 508 p.

8.II(ii). Periglacial Morphometry

BARBARA A. KENNEDY

Department of Geography, Cambridge University

1. Introduction

It is virtually impossible to find a universally acceptable definition of the term 'periglacial': for example, should it be restricted to the description of glacier-margin zones, or would this be too pedantic? The present discussion will take a highly pragmatic view and equate 'periglacial' areas with those in which the ground at depth is perennially frozen: these are the permafrost zones. Such a definition will include some high-altitude areas of frozen ground, in addition to the main high-latitude regions, but will exclude alpine meadows where the ground is completely thawed each summer. It should be stressed that the major characteristic of such periglacial regions is not the number of freeze–thaw cycles but the depth of frost penetration and length of time that the surface temperatures are below 32° F.

Although calculations have been made of the fraction of the world's water at present 'stored' in glaciers and ice-caps, no such estimates exist for that proportion which is removed from the hydrologic cycle as frozen ground water in regions of continuous and discontinuous permafrost. Figure 8.II(ii).1 shows the relative extent of ice-sheets and permafrost for the northern hemisphere (which contains the major areas of frozen ground), and it is clear that the volume of water involved in the latter regions is considerable.

With some exceptions (notably in the north-eastern U.S.S.R.) most of the areas of present periglacial climate were covered by continental ice sheets during the Pleistocene (see fig. 8.II(ii).1), and many of their landforms are therefore those common to all regions of glacial erosion and deposition. In addition, many coastal areas of frozen ground show clear effects of recent glacial 'rebound' in the form of flights of strandlines: the eastern coast of Baffin Island is a case in point.

Regarding other landforms, the opinion of geomorphologists is sharply divided. Peltier [1950, p. 221] has proposed a distinct periglacial cycle, resulting from the action of 'intense frost shattering, solifluction and congeliturbation'. The continued operation of these processes – upon an area initially possessing strong relief – is considered to result in 'extensive surfaces of cryoplanation with slopes less than 5°' which grade laterally into even lower-angle 'congelifractate-covered surfaces of downwastage or lateral planation' (Peltier [1950], p. 225).

Some workers have accepted the view of a distinct, periglacial assemblage of landforms: for example, Suslov [1961, p. 137] considers that the presence of

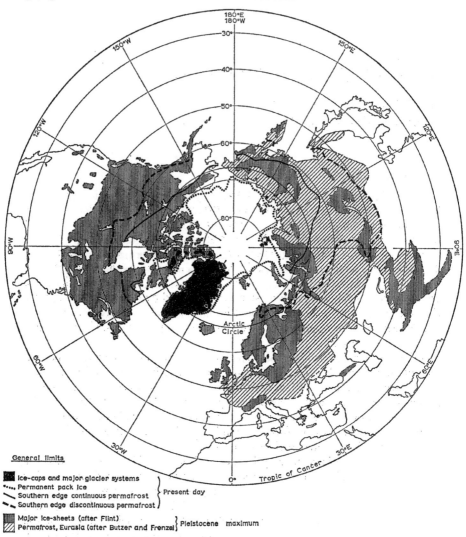

Fig. 8.11(ii).1 Past and present distribution of ice sheets and frozen ground in the northern hemisphere. 'Pleistocene maximum' should read 'Last glacial maximum'.
(The differences between some of these distributions and those shown in Fig. 8.1.1 reflect current differences of opinion.)

permafrost will affect the entire character of relief, though one should emphasize that this is in relation to eastern Siberia, an area of comparatively low relief which was not entirely glaciated. In much of northern North America, on the other hand, workers have felt it necessary to stress how little the landscape corresponds to the Peltier ideal (Bird [1953], p. 36; Mackay [1958], p. 26).

Rather than attempt to reconcile these views, the divergence of which may well result from historical differences between areas, the present discussion of periglacial landforms will be limited to those features which are uniquely present in regions of frozen ground. With the exceptions outlined below, the landforms of periglacial areas tend, perhaps disconcertingly, to resemble those of the other major fluvial regions of the earth's surface, both in their complexity and types. If one considers the vast extent of permafrost, as outlined in fig. 8.II(ii).1, and the range of climatic regimes, lithology, and relief to be found in such areas, the diversity of landforms is scarcely surprising.

2. Individual landforms

A. Pingos

'Pingo' has become the generally used term to denote certain ice-cored mounds or 'hydro-laccoliths' which are common in some low-lying permafrost areas of predominately fine-grained sediments.

A major concentration of pingos is found around the mouth of the Mackenzie Delta, N.W.T., Canada, where they have been intensively studied by Mackay [1963, pp. 69–94].

In form, pingos are conical hills of between 10 and 150 ft in height, with basal diameters ranging from 100 to 2,000 ft. The maximum heights are attained by those pingos with intermediate diameters between 500 and 700 ft, as are the steepest side slopes (up to 45°).

The occurrence of pingos is very closely related to the presence of drained lake-beds and marsh-filled channels, and it appears that the central ice-lens is formed, as shown in fig. 8.II(ii).2, by the migration of water as the permafrost front advances inwards from the old shorelines. The classification into 'closed' or 'open-system' pingos is related to the absence or presence of a continuing source of water after the formation of the initial core.

Although many pingos are long-established features of permafrost landscapes, all collapse in time; either because a change in climate destroys the permafrost altogether or because the over-lying sediments are cracked and the ice-core exposed to insolation. A collapsed pingo will be represented by an almost circular depression with a raised rim.

In themselves, pingos can scarcely be termed 'significant' landforms, yet in those permafrost areas in which they are found they create the nearest thing to a unique periglacial landscape.

B. Ground-ice slumps

Where sections of permafrost underlain by thick layers of ice are exposed to direct lateral undercutting – as along a river bank or sea coast – severe slumping ensues.

As the ice content of the ground may be anything from 50% upwards, the slump mass is liable to extreme wastage, and the chief feature which distinguishes these forms from slumps of other kinds is that the debris 'toe' is markedly

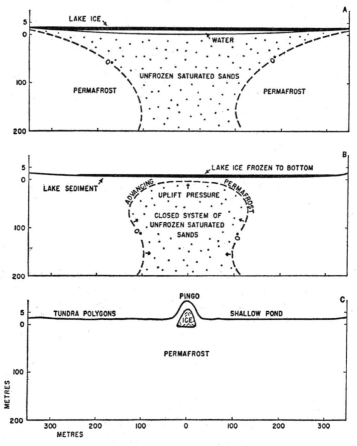

Fig. 8.II(ii).2 Schematic origin of a pingo (From Mackay, 1963).

In diagrams A, B, and C a vertical exaggeration of 5 × has been used for the height above zero in order to show the lake ice and the open pool of water.

(A) A broad shallow lake has an open pool of water in winter with a frozen annulus around it. No permafrost lies beneath the centre of the lake.

(B) Prolonged shoaling has caused the lake ice to freeze to the bottom in winter and induced downward aggradation of permafrost. Infilling has raised the lake bottom a small distance. The deepest part of the lake, which has the thinnest permafrost, is gradually domed up to relieve the hydrostatic pressure.

(C) The pingo ice-core, being within permafrost, is a stable feature. The old lake bottom is occupied by tundra polygons and shallow ponds. Because of scale changes in the diagram, the volume of the ice-core should not be construed as showing a direct relationship to the initial volume of unfrozen material.

underfit when considered in terms of the volume of the slump scar. The slumps themselves are arcuate-headed, and their sides may be bordered by levées of hardened mud.

C. Oriented lakes

Two of the three major occurrences of oriented lakes in North America are found in permafrost areas: near Point Barrow, Alaska, and in the Mackenzie Delta, N.W.T. (the third region is that of the Carolina 'bays'). The critical factor governing the development of such lakes in the North at the present time appears to be the strength of winds along the Arctic coast and the lack of barriers in the tundra landscape which will break their fetch.

In form, oriented lakes are elliptical in outline, with a triangular 'deep' at their centre. Lengths of the long axes may range from 100 ft to 2 miles, and the most common length:width ratio is 2:1 (Mackay, 1963).

The observation that such lakes tend to be oriented north–south, at right angles to the prevailing winds, has been shown to agree with forms derived from theoretical calculations of circulation patterns in two dimensions.

D. Ice-wedge polygons

These polygons are found in many low-lying areas of permafrost and are formed by the junction of ice wedges which initiate in tension cracks developed in the surface materials. In size, ice-wedge polygons may range from 5 to 100 ft across (fig. 8.11(ii).3), and they may possess either high or low centres, depending upon their stage of development.

E. 'Naleds' or icings

The build-up of hydrostatic pressure in an unfrozen aquifer in permafrost may become so great that water bursts through the overlying beds and freezes on the ground surface. Such icings are, obviously, winter features, but on melting they leave devastated areas comparable to those created by the passage of small avalanches.

3. Valley-side features

Although it is impossible to distinguish a 'periglacial valley form', some valleys within permafrost areas may possess minor characteristics which can be specifically related to the climatic regime.

A. Solifluction features

The process of solifluction – or downhill creep of soil with a very high water and ice content above the permafrost table – has frequently been cited as a dominant and characteristic feature of periglacial regions, but within any one area the evidence for its operation is usually given by very minor features of the landscape.

Among the forms attributable to solifluction are low, arcuate lobes and terraces and soil or vegetation stripes.

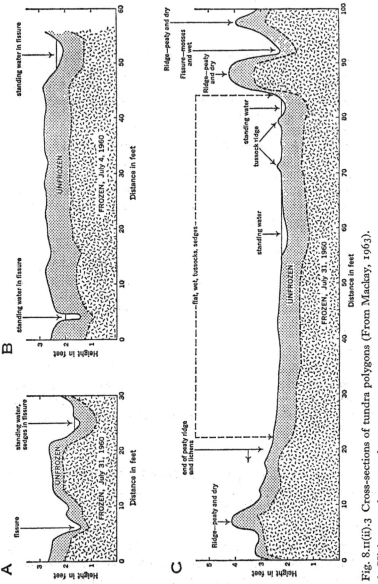

Fig. 8.II(ii).3 Cross-sections of tundra polygons (From Mackay, 1963).

A. High-centred peaty polygon.
B. Polygon subject to periodic inundation.
C. Low-centred polygon.

B. Nivation and snow-bank hollows

These concave niches in valley sides are not restricted to permafrost areas, but they are probably more common in such regions than elsewhere.

In size and appearance, nivation hollows may vary considerably. In some cases the back walls are bare, whereas the footslope, which may be vegetated, is rilled and frequently grades into a solifluction terrace downslope. In other cases the whole niche may be vegetated, giving no direct evidence of the processes at work.

The influence of the presence of such hollows upon slope form is also various. In most areas it would seem that profiles on which niches develop are steepened, but the degree of steepening depends upon the position of the niche on the profile and the proportion of the total length which it occupies. In areas of very low precipitation – notably Banks Island, N.W.T. – there is evidence to suggest that strong development of nivation hollows leads to a general decline in the angle of the slope above: possibly this relationship arises from the unusual concentration of moisture represented by the snow patch, which provides favourable conditions for solifluction on the back wall. Whatever the mechanism, slopes in the Kellet drainage, Banks Island, which are the site of persistent snow banks are, on average, 8° less steep than those without marked nivation hollows.

The most pronounced nivation hollows are found on north- or north-east-facing slopes, where insolation is weakest and late-lying snow banks consequently favoured.

C. Asymmetrical valleys

There is evidence to suggest that the steepening of south- and west-facing slopes, unrelated to structural controls, is peculiar to certain permafrost environments, though not all asymmetric valleys currently developing in such areas are of this type.

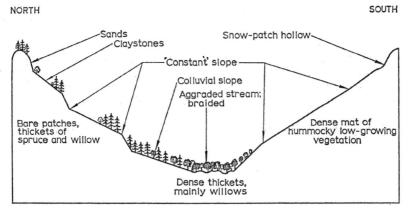

Fig. 8.II(ii).4 Cross-section of a typical asymmetric valley, Caribou Hills, North West Territories, Canada (From Kennedy and Melton, 1967).

Figure 8.11(ii).4 shows an idealized cross-section of a typically asymmetrical valley at the foot of the Caribou Hills, N.W.T. In this area south-facing valley sides have average maximum angles of 28° and are significantly steeper than the comparable sections of north-facing profiles, which average only 23°. The development of this asymmetry is strictly limited to the lower valleys, which have a relative relief of up to 200 ft and may reach a quarter of a mile in width. In the narrow, relatively shallow upper valleys north-facing slopes are significantly the steeper.

It appears that the development of periglacial asymmetry, in this sense, is controlled fairly strictly by valley dimensions. Only in deep, wide valleys does the combination of maximum insolation, greatest depth of active layer, and large downslope gravity component on south- and west-facing slopes lead to rapid, large-scale movement of surface material and steepening of the valley sides.

In narrower valleys, or those less incised, the steepening of north- and east-facing profiles is jointly controlled by the presence of large snow-patch hollows and slumping along the plane of the relatively shallow permafrost table, as the result of basal undercutting by streams.

REFERENCES

BIRD, J. B. [1953], *Southampton Island;* Memoir 1, Geographical Branch, Department of Mines and Technical Surveys (Ottawa), 84 p.

BIRD, J. B. [1967], *The Physiography of Arctic Canada* (Baltimore), 336 p.

BROWN, R. J. E. [1960], The distribution of permafrost and its relation to air temperatures in Canada and the U.S.S.R.; *Arctic,* **13,** 163–77.

BUTZER, K. [1964], *Environment and Archaeology* (London), 524 p.

FLINT, R. [1957], *Glacial and Pleistocene Geology* (New York), 553 p.

FRENZEL, B. [1968], The Pleistocene vegetation of northern Eurasia; *Science,* **161,** 637–49.

KENNEDY, B. A. and MELTON, M. A. [1967], Stream-valley asymmetry in an arctic environment; *Research Paper 42, Arctic Institute of North America,* Montreal, 41 p.

MACKAY, J. R. [1958], *The Anderson River Map – Area, N.W.T.;* Memoir 5, Geographical Branch, Department of Mines and Technical Surveys, (Ottawa), 137 p.

MACKAY, J. R. [1963], *The Mackenzie Delta Area, N.W.T.;* Memoir 8, Geographical Branch, Department of Mines and Technical Surveys, (Ottawa), 202 p.

PELTIER, L. C. [1950], The Geographical Cycle in Periglacial Regions; *Annals of the Association of American Geographers,* **40,** 214–36.

SUSLOV, S. P. [1961], *Physical Geography of Asiatic Russia;* Translated by Gershevsky, N. D., from the Russian 2nd edition of 1956, (San Francisco and London), 594 p.

9.II. Relation of Morphometry to Runoff Frequency

G. H. DURY

Department of Geography, University of Sydney

By strict definition, channel morphometry is the measurement of channels, while channel morphology is the study of channel shape. By extension, both terms are used to connote shape characteristics, whether in plan or in the vertical plane. Before the relation of morphometry to discharge frequency can be examined, it is necessary to define in this context the terms *pattern* and *habit*.

1. Channel pattern

This is the trace of a channel in plan, as shown, for instance, on vertical air photographs or as represented on maps. Meandering patterns and braided patterns occur widely. *Habit* in this sense can be substituted for *pattern*, as when a stream is said to have a meandering or a braided habit; but the study of habit as a mode of stream behaviour goes beyond the two dimensions of the plan view. It involves the shape of the channel in cross-section, and also the long-profile of the bed.

No comprehensive classification of channel pattern has yet been produced. The following provisional classification, including a number of well-represented types of which some are illustrated in fig. 9.II.1, is most unlikely to be complete:

1. Meandering
2. Braided
3. Straight
4. Straight-simulating
5. Deltaic-distributary
6. Anabranching
7. Reticulate
8. Irregular

Morphologic studies of channels have hitherto been concentrated on natural channels which are meandering, braided, or straight-simulating, while laboratory work has dealt with meandering, braided, and straight channels. These types will be chiefly discussed in what follows, meandering channels being first used to explain some of the principles involved.

Meanders are sinuous bends. In practice, a complete range of intermediate patterns links straight channels to channels which are highly sinuous and obviously meandering, while a second or possibly a variant continuum reaches

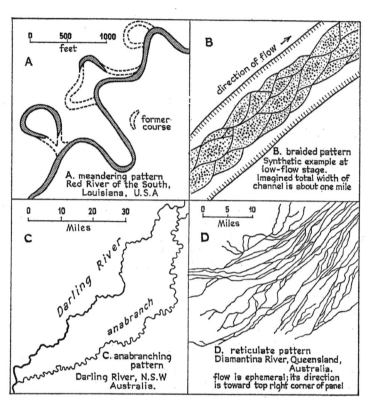

Fig. 9.11.1 Four widely represented types of channel pattern.

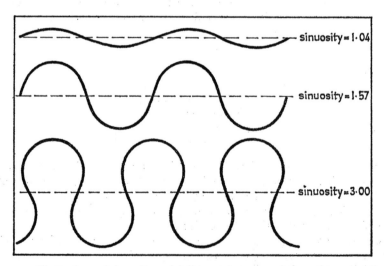

Fig. 9.11.2 Illustrations of sinuosity: axial distance is measured along the pecked lines.

from straight channels to channels which are merely irregular in plan. If it is required to discriminate between channels which meander and other single channels which do not some arbitrary fix on the scale of sinuosity is normally used. Sinuosity is the ratio of channel distance to axial distance (fig. 9.11.2): a sinuosity of at least 1·5 is used by some workers as a criterion of meandering.

In addition to sinuosity, the attributes of a meandering channel include amplitude, wavelength, and radius (fig. 9.11.3). Amplitude, which at one time commanded a great deal of attention, is not greatly considered nowadays. Radius is a somewhat crude measure of channel curvature, since meander bends tend to assume the form of sine-based waves rather than that of arcs of circles. Wavelength is fundamentally significant, especially in relation to bedwidth and

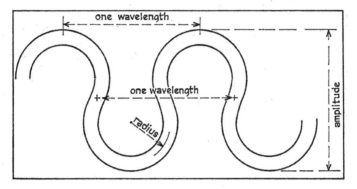

Fig. 9.11.3 Some elements of meander geometry.

to discharge. Variations in the width/depth ratio ensure that the wavelength/width ratio is also variable; but on many streams the wavelength is between 8 and 12 times the bedwidth, with the modal value between 9 and 10 times bedwidth.

A meandering channel is asymmetric in cross-section at bends, the greatest depths occurring near the outer banks. Between bends it is more symmetrical, and shallower. The deeps at bends are *pools*, while the shallows between bends are *riffles*. Meandering appears to begin with the establishment of a pool-and-riffle sequence. Straight laboratory channels, shaped in homogeneous materials, can deform into pool-and-riffle when water is fed straight into them at constant discharge. We infer that straight uniform channels are unstable. A characteristic spacing from one pool to the next in a flume experiment is about five bedwidths. As meanders form, alternate pools migrate to alternate sides, giving the approximate wavelength of two inter-pool spacings, or 10 bedwidths, as in nature (fig. 9.11.4).

Field experiments with marked sediments show that riffles tend to be eroded somewhat at low stages, the material removed from them being fed into the pools. When discharge increases and stage rises to banktop or thereabouts the pools are scoured anew, the sediment excavated from them lodging on riffles farther downstream. That is, the fullest expression of the pool-and-riffle sequence is promoted by discharge at or near bankfull. This circumstance reinforces what was said in an earlier section about channel-forming discharge.

Bedwidth on meandering channels, and on single channels generally, is accordingly measured between banktops, unless water-surface width is specific-

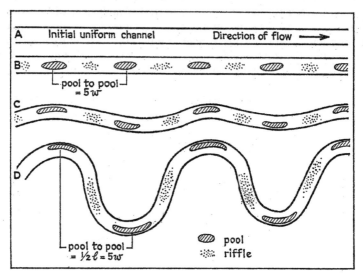

Fig. 9.11.4 A, uniform channel, no pool-and-riffle; B, straight channel with pool-and-riffles; C, slightly sinuous channel; D, meandering channel.

ally in question. A well-established relationship exists between bedwidth, w, and bankfull discharge, q_{bf}, in the form

$$w \alpha q_{bf}^{0.5}$$

or, numerically, with w in feet and q in ft³/sec, approximately

$$w = 3q_{bf}^{0.5}$$

Meander wavelength, l, has already been described as an approximate linear function of bedwidth, i.e.

$$l = kw$$

or, numerically,

$$l = 10w$$

Independent analysis produces the expectable result that

$$l \alpha q_{bf}^{0.5}$$

or, in the same units as above,

$$l = 30q_{bf}^{0.5}$$

The values of the numerical constants are affected in specific cases by variation of the width/depth ratio of the channel.

The connection between channel size and meander size (wavelength), on the one hand, with value of discharge, on the other, takes this discussion a long way from the outmoded proposition that meandering is caused by obstacles. Precisely the opposite is true. Obstacles, including variation in the cohesiveness of alluvium, distort and even suppress meanders, as is well shown by studies of the lower Mississippi. Meander patterns can be produced on a cathode-ray screen by means of electrical analogue devices; when suitable resistances are introduced into the circuits distorted bends and cut-offs appear which can be matched from the traces of actual rivers.

Just as the development of pool-and-riffle indicates that a uniform bed-slope is unstable, so does the side-to-side swing which produces meanders indicate that a straight trace is unstable. It can be proved theoretically that if a channel deviates from straightness a meandering trace is in the statistical sense the most probable trace. It minimizes the variability of water-surface slope downstream, the variability of bed shear, and the variability of friction. Hence, once a meandering pattern has been established, it is likely to persist, unless some really powerful disturbing factor comes into operation. But although a great deal is known about the behaviour of meandering streams and about their statistical properties, it is still not possible to define the ultimate cause of meandering.

A number of penetrating descriptive and analytic studies of braided channels are on record, although not a great deal has been done in relation to channel-forming discharge. The braided pattern is best displayed at low stages, when the characteristic multiple bars are revealed, interlaced by the dividing and recombining minor channels. At common low stages the bars may cover, perhaps, 80% of the streambed.

Braided channels have much greater width/depth ratios than have meandering or other single channels. Thus, the highest flow velocities, which occur close to the water surface, are also close to the bed in braided channels. This is to say that the gradient of velocity from surface to bed is steep, and that shear stress on the bed is powerful. Braided patterns are associated with mobile streambeds, which are typically deformed into the roughly diamond-shaped bars which separate the minor low-stage channels. Given mobile bed material of the sand grade or coarser, anything which promotes widening and associated shallowing of the channel is likely also to promote braiding.

One possible cause of widening, shallowing, and braiding is weakness of the banks. If these are incohesive, collapsing so rapidly that they impel the channel to become wider and shallower, braiding can result. A variant on this situation is provided by outwash streams of glacial meltwater, which, because they are filling in the valleys that they occupy, are virtually without banks in the ordinary sense. At the opposite extreme come channels with a high proportion of fine (silt-clay) material in their sediment, where the width/depth ratio is low and where braiding is improbable.

A particularly illuminating case of braiding is that where a meandering stream converts to the braided habit, either in a given reach or in the downstream direction. It can do this if its floodplain alluvium is underlain by coarser material, such as glacial outwash or older alluvium deposited by a former larger stream, and if one or more pools are especially deep at unusually sharp distorted bends. The bar thrown up immediately downstream of a deep pool and sharp bend splits the channel into two. Each of the divided channels is less efficient than the single channel which it partly replaces, simply because it is smaller. The loss of efficiency can be compensated by a steepening of slope: the stream deepens

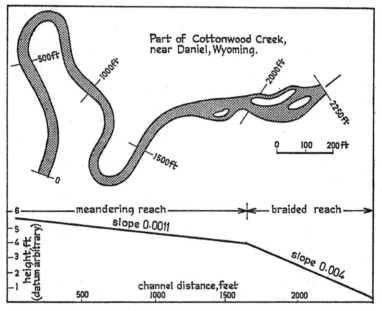

Fig. 9.II.5 An example of changing pattern and changing slope (Adapted from Leopold and Wolman, 1957).

each of the two lesser channels, helping the bar to dry out at most stages and permitting it to be colonized by plants. Repeated subdivision in this fashion can convert a meandering into a braided trace.

Because of the inefficiency of its total channel, a braided stream will have a steeper slope than a meandering stream of equal channel-forming discharge (fig. 9.II.5).

Truly straight channels scarcely exist outside the laboratory, except where a headwater stream is firmly held on the line of a fault. Straight-simulating channels are discussed below. The class of streams with irregular traces is a ragbag into which is stuffed patterns not otherwise classified. Deltaic-distributary streams, occurring in the special conditions indicated by their title, will not be further discussed here, except for the comment that some are notably meandering, as in certain deltas of South-East Asia, while others, including some of the Mississippi passes, are just as notably straight.

Anabranching and reticulate channel patterns are widely represented on the pediplains and alluviated basins of inland Australia, but are unlikely to be confined to these localities. Work on anabranching channels has been restricted mainly to the alluvial plains of the Murray and Murrumbidgee, where the anabranches – offshoots – rejoin the original trunk or unite with a next-neighbouring trunk, sometimes after a distance of tens of miles. Little rigorous work has yet been done on reticulate channels, which resemble braided channels in being complexly subdivided at low stages; but they differ from braided channels in being typified by markedly ephemeral flow, and in having a distinctly open network. They occupy whole valley floors as opposed to channel beds. Anabranching and reticulation alike seem to be responses to very gentle transverse slopes, which allow the current to subdivide freely at times of high stage. Both kinds of pattern are widely enough represented for them to constitute distinct classes.

2. Floodplains

By definition, floodplains are liable to inundation. They are best understood in the context of meandering streams. Shifting meanders work over the valley-bottom alluvium, eroding their outside bends but depositing on the insides and forming a depositional strip along the length of the valley. Some floodplains are composed chiefly of point-bar deposits laid down in multiple crescents on the inner sides of bends, with nil to minor additions of sediment from floodwaters, but others may be constructed mainly by overbank deposition.

Another kind of difference is that some floodplains possess levées – natural raised banks bordering the channel – whereas elsewhere levées are absent. The lower Mississippi is a noteworthy levée-builder, causing backswamps to form on its outer floodplain and obstructing the entry of tributaries (fig. 9.11.6). On very large streams such as this it is usual to find routeways, and even settlement, concentrated on the levées, out of reach of some floods and away from the ill-drained backswamps. The lower Thames, by contrast, like the Cotswold feeders of the upper Thames, is typically devoid of levées, with a floodplain sensibly flat in cross-section: on such a floodplain there is nothing to choose between one location and another, in respect of wetness or of freedom from flood danger.

Differences between channel-side accretion and overbank deposition, and between levée-building and the lack of it, are not yet fully understood, but appear to be related to the calibre of sediment transported through the channel, and to the relative proportion of bed load and suspended load. The reasoning involved in this interpretation can be used to classify braided channels as of the depositing bedload type.

If a meandering stream on a floodplain is cutting down very slowly or not at all it can be expected to reach bankfull stage once every year, and to inundate its floodplain in about two years in three. Many present-day meandering streams

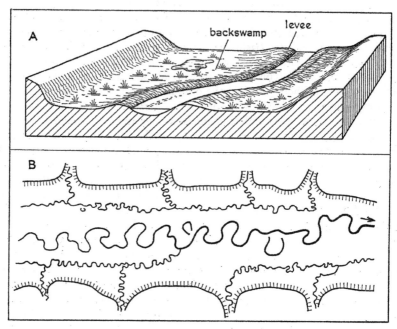

Fig. 9.11.6 A, levee–backswamp association; B, deferred tributary junctions.

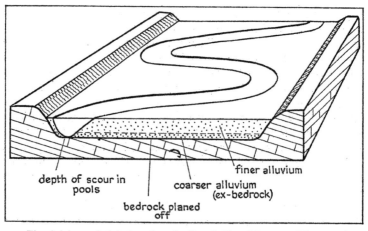

Fig. 9.11.7 Floodplain underlain by planed-off rock floor (Compare Fig. 9.11.9).

are, however, suspected to be slightly incised, so that their frequencies of bank-full flow and of inundation are reduced below the expectable level. Other streams, engaged in raising their floodplains, cause unusually frequent inundation.

If the working-over of the floodplain by shifting meanders keeps pace with any downcutting, then the base of the floodplain will correspond to the maximum depth of scour in pools (fig. 9.11.7). In actuality, the floodplain base is often cut across older alluvium rather than across bedrock, for reasons which will appear under the next subhead.

3. Meandering valleys

Large numbers of streams occupy meandering valleys, where steep crescentic slopes on the outsides of bends oppose gentler lobate slopes on the insides, but

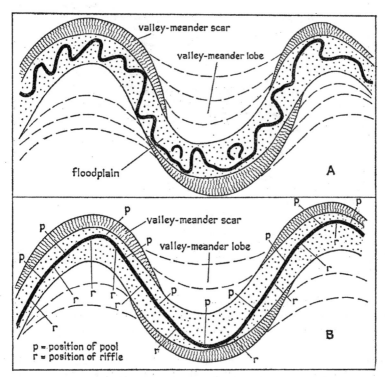

Fig. 9.11.8 A, manifestly underfit stream; B, underfit of the Osage type. Both portrayed as occupying ingrown meandering valleys.

where the stream channel is shaped wholly in alluvium and describes curves far smaller than those of the valley (fig. 9.11.8*a*). The steep outer slopes of the valley are called *meander scars*, the gentler inner slopes are *meander lobes, spurs,* or *slip-off slopes.*

The combination of scars and spurs belongs to trains of *in-grown meanders* which have increased in sinuosity during downcutting. The increase in sinuosity is not infrequently documented by crescentic patches of terrace on the slip-off slopes; and where the terraces are well enough preserved to reconstruct much of the sequence of downcutting it usually appears that the ingrowing stream was very little sinuous when it began to incise. The curves of the valley, that is to say, were produced largely or entirely during the course of incision.

Meandering streams on floodplains in the bottoms of meandering valleys are called *manifestly underfit*: manifestly because their condition is self-evident, and underfit because their bends are significantly smaller than the bends of the valley.

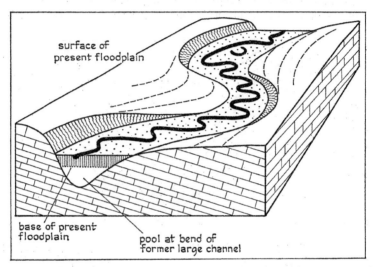

surface of
present floodplain

base of present
floodplain

pool at bend of
former large channel

Fig. 9.11.9 A common condition of manifestly underfit streams in ingrown meandering valleys: the fill beneath the base of the present floodplain is older than the floodplain alluvium.

The valley bends were cut when channel-forming discharge was far greater than it now is. On some manifestly underfit streams the channel of the large former stream is preserved beneath the present-day floodplain (fig. 9.11.9), the present-day floodplain constituting merely the topmost layer of fill.

According to location, the wavelength of valley bends ranges from about 5 times to about 10 times the wavelength of present-day stream meanders; the 5/1 ratio is widely represented. Calculation based on the connection $l \propto q_{bf}^{0.5}$ between meander wavelength and bankfull discharge gives a 25-fold increase of bankfull discharge for a 5-fold increase of wavelength. When allowance is made for additional hydraulic characteristics, such as channel size, channel shape, channel slope, and roughness, the requirement comes down to about a 20-fold increase in bankfull discharge.

Now although in special conditions underfitness may be due to river capture, the cessation of meltwater discharge from glaciers, or the cessation of overspill from ice-damned lakes, most streams which are now underfit have had their channel-forming discharges reduced by climatic change. It is highly likely that increase, reduction, and renewed increase of discharge occurred repeatedly during the Pleistocene, but little is known of any but the last main episode of shrinkage. This, possibly itself interrupted by an increase, took place between about 12,000 and 9,000 years ago. This was the time of the last major transition from high-glacial to interglacial conditions. When allowance is made for the reduced air temperatures of the time it can be shown that the swollen channel-forming discharge required to shape valley meanders could be produced by an increase in mean annual precipitation to $1\frac{1}{2}$ or 2 times its present value. Not only did the rise of temperature and the reduction of precipitation to present-day levels cause streams to become underfit: they also dried up the former pluvial lakes of arid lands.

Since the last main episode of stream shrinkage and channel filling, lesser fluctuations of temperature and precipitation have produced lesser episodes of renewed channelling and subsequent filling.

Not all meandering valleys are occupied by underfit meandering streams, a circumstance which until recent years has prevented the recognition of underfitness as general throughout vast areas. It is now known that a second type of underfit stream, the *Osage type* (fig. 9.11.8(*b*)) is quite common. This type is named after the Osage River of Missouri, U.S.A., where its qualities were first recognized. Although contained in a meandering valley, an underfit of the Osage type lacks stream meanders; but it possess a pool-and-riffle sequence, with the pools and riffles spaced appropriately to the existing channel at intervals of about 5 bedwidths. Because the existing channel is smaller than that of the former stream which cut the valley bends, the pool-and-riffle spacing is too close for these bends and for the curves which they impose on the existing channel: that is, pools occur more numerously than do channel curves. The stream has reduced its bedwidth and has correspondingly reduced its pool-and-riffle spacing, without going on to develop meanders. It behaves as if it were straight, apart from being enforcedly inflected round the curves of the valley. Thus, channels of underfits of the Osage type are assigned to the straight-simulating type of pattern.

The apparent wavelength/bedwidth ratio on streams of this kind is in many cases about 40/1, far larger than the rough 10/1 obtained for alluvial meandering channels; but in actuality the ratio on underfits of the Osage type is L/w, that between the wavelength of the large former stream and the bedwidth of the shrunken present-day stream. Whereas manifestly underfit streams can be recognized on sight, underfits of the Osage type can only be presumptively identified by a high apparent ratio of wavelength to width: their existence can only be proved (or disproved) by survey of the bed-profile and an investigation of the relationship between bedwidth and pool-and-riffle spacing.

REFERENCES

DURY, G. H. [1965], General theory of meandering valleys; *U.S. Geological Survey Professional Paper* 452. (Examines underfit streams.)

DURY, G. H. [1966], Incised valley meanders on the lower Colo River, New South Wales; *Australian Geographer*, **10**, 17–25. (Names the Osage type.)

LANGBEIN, W. B. and LEOPOLD, L. B. [1966], River meanders: Theory of minimum variance; *U.S. Geological Professional Paper* 422-H.

LEOPOLD, L. B. and WOLMAN, M. G. [1957], River channel patterns: Braided, meandering and straight; *U.S. Geological Survey Professional Paper* 282-B. (The basic reference on channel patterns.)

LEOPOLD, L. B. and LANGBEIN, W. B. [1966], River meanders; *Scientific American*, **15** (6), 60–9. (A simple, but valuable, account of meanders.)

LEOPOLD, L. B., WOLMAN, M. G. and MILLER, J. P. [1964], *Fluvial Processes in Geomorphology* (Freeman, San Francisco and London), 522 p. (This contains a very complete summary up to 1964.)

WOLMAN, M. G. and LEOPOLD, L. B. [1957], River flood plains: Some observations on their formation; *U.S. Geological Survey Professional Paper* 282-C.

10.II. Climatic Geomorphology

D. R. STODDART

Department of Geography, Cambridge University

The rise of climatic geomorphology dates from the period of exploration of the new tropical empires of France and Germany at the end of the nineteenth century: with its concern with the unusual and spectacular landform, the subject still bears the mark of this early work by scientific explorers. Prominent among these were Von Richthofen in China, Passarge, Jessen, Walther, and Thorbecke in Africa, and Sapper in Central America and Melanesia. This rapid expansion of knowledge of hitherto remote parts of the globe followed closely on the rediscovery in the 1860s and 1870s of fundamental geomorphic principles in the arid west of the United States, and their codification by Davis in the cycle-of-erosion concept from 1883 onwards.

In assimilating the new data into accepted theories, Davis treated the landforms of non-temperate climatic regions as deviants from the 'normal' scheme. In Germany each climatic region was thought to have its own assemblage of characteristic landforms and sequences of development. In more recent years German workers such as Büdel and Louis have continued to refine the concept of distinct morphoclimatic regions, while increasingly aware of the importance of climatic changes; while the French have viewed climate as but one, though a major, control of landscape morphology.

There are three central themes which require discussion in any treatment of climatic geomorphology. They are: (1) the view that landforms differ significantly betwen different climatic areas; (2) that these differences are the result of areal variations in climatic parameters and their effect on weathering and run-off; and (3) that, though considerable, climatic changes in Quaternary times have not disguised the climate–landform relationship. Each of these is attended by considerable difficulties, and can only be treated in broadest outline here.

1. Reality of distinctive climatogenetic landforms

While a general distinction has long been drawn between the landforms of arid, glacial, and humid temperate lands, surprisingly little objective morphometric evidence exists on landform variation within the wide range of fluvial conditions. Partly this results from the poor quality of topographical mapping over many parts of the earth's surface: Eyles [1966] has shown that Malayan maps, for example, are inadequate for many morphometric purposes. Hence morphometric work on maps often fails to demonstrate significant differences in landforms between climatically diverse areas. Viti Levu, Fiji, forms a good example: the south-east windward side of the island is wet, forested, and deeply dissected, with rainfalls often exceeding 3 m; the leeward side is under grassland, with rainfalls often less than 0·5 m. Measurements of drainage density, hypsometric integral, and Horton parameters for the Fiji 1:50,000 maps has failed to demonstrate significant differences in form between various rainfall and vegetation groupings. The Fiji maps are of good quality by comparison with many tropical areas.

Peltier [1962] adopted a less intensive approach, and sampled topographic

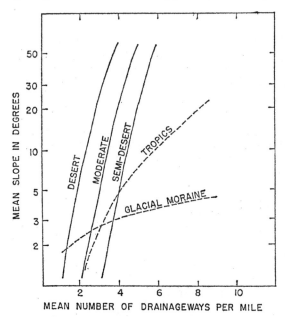

Fig. 10.11.1 Morphometry of major climatic regions (After Peltier, 1962).

maps on a world basis by randomly selecting geographical co-ordinates, correcting for latitude so that high-latitude areas were not over-represented. Because of the great variability in the scale of available maps, he measured the maximum difference in height within a 100-square-mile area at each sampling point, and used this to calculate mean relief in ft/sq. mile and mean slope in degrees. He then classified his sampling locations into major climatic groups (tundra, microthermal, mesothermal, and tropical), and also calculated the mean number of drainageways per mile at each station as a measure of topographic texture. When mean slope is plotted against mean number of drainageways per mile by climatic regions (fig. 10.11.1), the curves for desert, mesothermal, and microthermal areas are almost parallel, but with sideways displacement in response to changing runoff, whereas the curves for the tropics and for glacial morainic country are notably anomalous.

This suggests that first-order distinctions may be made between glaciated, tropical, and all other fluvial landscapes, and that second-order differences exist between desert, semi-desert, and temperate fluvial landscapes. The second-order differences could be interpreted in terms of Langbein and Schumm's sediment–yield curve, with erosion limited at lowest rainfalls by lack of water and at higher rainfalls by vegetation growth, reaching a maximum in the semi-arid lands. Two points need to be made here: first, the coarseness of Peltier's measures needs to be emphasized, for it raises the major problem of the scale at which climatic effects become apparent in landforms; and second, the fact that in Langbein and Schumm's scheme vegetation plays a major intermediary role between climate and landform.

Most identifications of climatically-controlled landscapes are made on two levels: (a) generalized impressions of whole landscapes, and (b) the recognition of peculiar landscape components. In the first case an illusion of differences in landform is certainly given by differences in the vegetation cover. In the Viti Levu case already mentioned the heavy green tropical forest of the wetter side, concealing the ground surface, contrasts with the brown and yellow appearance of the bare grasslands, where details of form are clearly apparent. On air photographs the forest would tend to conceal much fine dissection and the grass cover to reveal it, thus leading to underestimation of drainage density in the wet lands and to exaggeration in the dry. In the field, however, vegetation differences, reflecting climate, often suggest morphometric differences which may be un-critically accepted.

The recognition of peculiar landform components thought to be diagnostic of particular climates has been the chief tool of the climatic geomorphologist, in the absence of quantitative work on landform geometry. Several examples of such diagnostic landforms may be mentioned.

A. Surface duricrusts (Laterites, silcretes, and calcretes)

Laterites are often taken to indicate humid tropical conditions, though most laterite pavements are found in subtropical areas such as Bihar and the Guianas, and are frequently if not invariably related to past geomorphic conditions. Typically laterite pavements outcrop on interfluves between incised streams, under conditions of stripping of surface soil. Such crusts are often interpreted as of Tertiary age, or as having been under continuous formation since the end of the Mesozoic. Exposures of silcretes and calcretes similarly are often related to past rather than present climatic conditions.

B. Inselbergs

The term inselberg has been loosely used since Bornhardt's explorations in East Africa for steep-sided residual hills rising above low-angle plains in semi-arid areas, and Cotton has termed this assemblage the 'savanna landscape'. Inselbergs have been described over a wide variety of climatic conditions, from humid subtropical in Georgia, North America, to humid tropical in the Guinea coast-lands, south India, and Brazil, and to desert areas in western North America,

Mauretania, and south-west Africa. While these examples cover a wide variety of forms, it is possible to argue that inselbergs are lithologically or structurally controlled azonal features resulting from the combination of sharp areal variability in the resistance of rocks and of efficient debris-transporting mechanisms. Many inselbergs on resistant rocks are so large that they certainly predate Quaternary climatic changes: hence present climates are not necessarily those in which the inselbergs were formed.

C. Pediments

Pediments as classically described are smooth low-angle slopes surrounding desert mountains, formed by processes of either sheet-flooding or lateral stream migration as transportational or erosional successors of back-wearing mountain slopes. The controversy over the origin of pediments has tended to obscure their diversity of form, to which Tuan drew attention in south-east Arizona. The model of pediment formation by slope retreat was remarkably similar to one of Penck's models of slope evolution, which Penck himself believed to be tectonically rather than climatically controlled, and later workers, such as Lester King, have suggested that, far from being diagnostic of aridity and thus atypical, the pedimentation process is in fact the universal mode of slope evolution. Davis himself, in his 1930 paper on 'Rock floors in arid and humid climates', came close to this position. Many arid-zone pediments are clearly polycyclic, developed during the complex sequence of Pleistocene pluvials and interpluvials: many appear to be being destroyed under present climatic conditions, rather than being formed by them.

D. Tors

Tors resemble inselbergs in many respects apart from scale. In Britain Linton has interpreted the Dartmoor tors as indicative of Tertiary deep tropical weathering of an inhomogeneous rock, with subsequent stripping of the weathered mantle to reveal piles of corestones. Palmer and Nielson showed the importance in this case of Pleistocene frost-shattering and mass-movement under periglacial conditions; but no one would suggest that the morphologically similar features of, for example, eastern Nicaragua and Rhodesia originated periglacially.

In each of these cases of the climatic interpretation of distinctive landforms two conclusions stand out: first, that form is an ambiguous guide to origin, and is often more complex than first generalizations would suggest, and second, that most supposedly diagnostic forms are older than Pleistocene climatic fluctuations and cannot be assumed to be genetically related to the climates in which they are now found. Landscapes differ, of course, in the degree of genetic ambiguity attached to them: desert dunes are quite clearly related to arid conditions, and even coastal dunes are largely absent in the humid tropics. Fossil dunes have been successfully used by Grove and others to reconstruct the former extent of African deserts in the Pleistocene. At the opposite end of the ambiguity scale, by contrast, are such forms as summit convexities on slopes (interpreted both paleoclimatically and structurally by different workers in Arizona) and V-shaped and saucer-shaped river valleys, described by Louis as typical of different climatic zones in Tanzania but found adjacent to each other in the New Guinea uplands by Bik.

Though landscapes often look different in different parts of the world, we must conclude that, with the exceptions of extreme cases, such as the glacial and arid, landscape assemblages of other climatic zones have yet to be unequivocally interpreted in climatic terms, and that many of the type-landforms of particular climatic zones are themselves of dubious significance. This is not to deny that climatically controlled landform differences exist, though morphometric confirmation of this is scanty; but it is to assert that the climatic inputs and geomorphic outputs in denudation systems are so little known that one cannot be inferred from the other.

2. Climatic controls of geomorphic significance

Because of the dearth of morphometric data, most treatments of climatic geomorphology concentrate either on the climatic characteristics associated with particular landscape types or on the definition of climatic zones within which

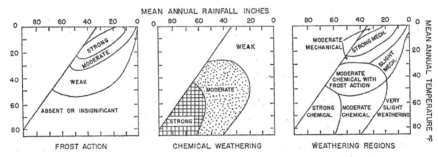

Fig. 10.11.2 Climatic control of frost action, chemical weathering, and weathering regions (After Peltier, 1950).

distinct landform assemblages might be expected to develop. Peltier [1950], in the best-known scheme of morphogenetic regions, distinguishes nine, of which six have been generally recognized in the literature (glacial, periglacial, 'moderate', savanna, semi-arid, arid) and three have yet to be worked out (boreal, maritime, selva). W. M. Davis had recognized only the 'normal', arid, and glacial cycles, and in spite of working in the humid tropics, he did not add a separate scheme for low latitudes.

Peltier's scheme (Table 10.11.1) uses two climatic parameters, mean annual temperature and mean annual rainfall: the morphogenetic regions are defined in terms of dominant processes varying areally and in combination, and not in terms of landform geometry (figs. 10.11.2 and 10.11.3). It is surprising that few attempts have been made to refine this scheme in twenty years. Apart from wind action in arid lands and ice action in cold ones, regions are differentiated in terms of water availability, both in channels and on slopes. A first refinement would thus be to replace mean annual rainfall by a measure of the availability of water for geomorphic work, for example, in crude terms, rainfall less evapotranspiration, or rainfall less potential evaporation. Calculations involving these measures have been made by Chorley and by Tanner. The more refined the attempts to give precision to climatic limits, however, the greater the problem of what the limits are for, and the more acute the difficulties of local variability. Standard meteorological measures are not necessarily those of greatest geomorphic significance, and Visher has experimented in mapping a range of climatic parameters of presumed geomorphic value. Many of these are not simply additive in their effects, and the problem of generalizing morphogenetic regions from them

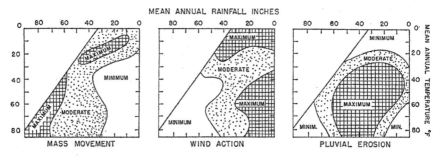

Fig. 10.II.3 Effectiveness of mass movement, wind action, and pluvial erosion under different climatic conditions (After Peltier, 1950).

remains. The problems of delimitation using more complex parameters are illustrated by the problems of the boundaries of the humid tropics: using temperature and atmospheric humidity criteria, for example, Garnier has mapped as parts of the humid tropics areas such as southern Arabia, which on vegetational criteria are manifestly not.

The climate at the ground, and even more the climate in the soil, normally differ considerably from the climate of the Stevenson screen. Thus in the humid tropics the interposition of a 30–50-m layer of vegetation between atmosphere and lithosphere means that the climate of the open air affects geomorphic processes only indirectly. Detailed studies of forest climate have been made in Uganda, Nigeria, Colombia, and the Guianas. Roughly two-thirds of the rainfall reaches the ground as rainfall: the rest is intercepted by leaves, evaporated, or channelled down trunks. The finest rains are filtered out and may not reach the ground at all, and the mean droplet size is markedly increased. Temperature and humidity variations are greatly reduced under the forest canopy, and in the soil temperature is invariant at quite small depths. Direct insolation at the surface is replaced by complex patterns of sunfleck and shadow. Comparative data on ground climate are available from many temperate forests, but are lacking for many areas (Geiger, 1965).

Data are simply not available to map the distributions of these geomorphic climates, and thus we generally substitute other distributions. The composition and distribution of vegetation was used by Köppen in his search for significant climatic boundaries, and vegetation is a major criterion in the morphoclimatic maps of Tricart and Cailleux. Landform, vegetation, and climate are, of course, complexly interrelated, and it is questionable whether vegetation is not as ambiguous as other measures in the search for meaningful criteria.

TABLE 10.11.1 Morphogenetic regions

Morphogenetic region	Estimated range of average annual temperature (° F)	Estimated range of average annual rainfall (in.)	Morphologic characteristics
Glacial	0–20	0–45	Glacial erosion Nivation Wind action
Periglacial	5–30	5–55	Strong mass movement Moderate to strong wind action Weak effect of running water
Boreal	15–38	10–60	Moderate frost action Moderate to slight wind action Moderate effect of running water
Maritime	35–70	50–75	Strong mass movement Moderate to strong action of running water
Selva	60–85	55–90	Strong mass movement Slight effect of slope wash No wind action
Moderate	38–85	35–60	Maximum effect of running water Moderate mass movement Slight frost action in colder parts No significant wind action except on coasts
Savanna	10–85	25–50	Strong to weak action of running water Moderate wind action
Semi-arid	38–85	10–25	Strong wind action Moderate to strong action of running water
Arid	55–85	0–15	Strong wind action Slight action of running water and mass movement

Source: Peltier [1950], 215.

Recognition of the interdependence of these controls leads, however, to important insights, which are often most readily apparent when the system is deranged. This is well demonstrated in the humid tropics, in the comparison between erosion under forest, and on bare ground, or between forests and savanna. Rougerie found sediment yields from experimental plots in the Ivory Coast to be roughly fifty times greater from bare ground than under forest. The effects of forest clearing during shifting cultivation, or of forest replacement by

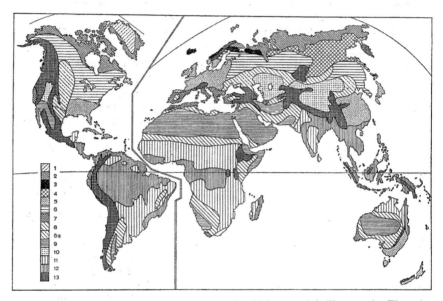

Fig. 10.11.4 World morphoclimatic regions (After Tricart and Cailleux, 1965, Fig. 49).

1. Glaciated regions.
2. Periglacial regions with permafrost.
3. Periglacial regions without permafrost.
4. Forest on Quaternary permafrost.
5. Mid-latitude forests: with maritime climate or lacking severe winter.
6. Mid-latitude forests: with severe winter.
7. Mid-latitude forests: Mediterranean type.
8. Semi-arid steppes and grasslands.
8a. Semi-arid steppes and grasslands: with severe winter.
9. Deserts and degraded steppes: without severe winters.
10. Deserts and degraded steppes: with severe winters.
11. Savannas.
12. Intertropical forests.
13. Azonal mountainous regions.

savanna grassland, either caused by man's activities or by climatic changes, would thus be of major geomorphic importance. Rougerie also found that a simple sediment–yield curve relating erosion and rainfall could not be constructed, for the amount eroded by a given rainfall varied with antecedent conditions, both seasonal and short-term. Rains at the beginning of the wet season gave higher sediment yields than later rains, though different trends were apparent with different rainfall intensities.

Rougerie's work highlights another major problem in defining climatic parameters: that of periodicities and magnitudes. In simplistic terms it is possible to draw contrasts between, for example, the sparse but occasionally torrential rainfalls of the deserts, the strongly seasonal and high-intensity rainfalls of the savannas, and the continuous high-intensity rains of the humid tropics, and to infer geomorphic consequences from them. But the more we learn of rainfall characteristics, the less they appear to conform to the assumed model (Beckinsale, 1957; Peel, 1966), and it is doubtful whether any useful purpose is served by elaborate deductive reasoning about what might be happening.

On a world scale, therefore, it is possible to describe the distribution of certain attributes of climate of geomorphic importance, such as thunderstorm incidence,

TABLE 10.11.2 Tricart and Cailleux's morphoclimatic zones

---

1. Cold zone
    (a) Glacial domain
    (b) Periglacial domain

2. Forested zone of middle latitudes (modified by man, with glacial and periglacial survivals):
    (a) Maritime domain, mild winter (strong survival of glacial and periglacial forms).
    (b) Continental domain, severe winter (Quaternary permafrost may survive).
    (c) Mediterranean domain, dry summers.

3. Arid and sub-arid zone of low and middle latitudes:
    (a) Rainfall distinction: steppe and desert.
    (b) Winter temperature distinction into cold and warm regions.

4. Intertropical zone, differentiated by rainfall seasonality:
    (a) Savannas.
    (b) Forests.

Source: Tricart and Cailleux [1965, pp. 268–88].

frost frequency, and tropical cyclones (Common 1966); it is also possible to map the zonal distributions of soils and of vegetation. Several workers have combined such criteria with those of relief to produce maps of morphoclimatic regions, in which the critical though only inferred parameters must be weathering-climate and availability of water for transportation of debris. Table 10.11.2 lists the morphoclimatic regions recognized by Tricart and Cailleux [1965, pp. 268–88], and fig. 10.11.4 shows their distribution. Apart from the arid and glacial lands, such maps do not record the occurrence of different kinds of processes, only the variety of combinations of the same processes: the differences are of degree and not of kind. Further, it does not follow that in such maps we are also mapping the boundaries of landform systems, many of which may not be related to present climates.

3. The problem of climatic change

Since the time of Agassiz, evidence has accumulated that not only has climate changed drastically and rapidly over large parts of the earth during the Quaternary but stratigraphic evidence shows that climatic changes have occurred during most of geological time. These changes pose a critical problem in climatic geomorphology, for it cannot be assumed that the landforms found in any given climate have developed in response to it. How, then, can the links between climate and landform be identified?

Albrecht Penck was aware of this problem, but believed that the glacial shifts of the climatic belts were minor, with changes limited to marginal zones. In the central parts of the deserts the humid tropics and other zones he considered Pleistocene changes to have been minimal. It is true that the most spectacular evidence of recent climatic change comes from the margins of the deserts, where relict dune fields now vegetated indicate former dry conditions, and old shorelines and lake deposits wetter phases. Thus Grove has described an ancient erg in Hausaland, Nigeria, and evidence for a 'Mega-Chad' which may have existed only 10,000 years ago. Comparable forms are known from the western United States and from Australia. The sensitivity of the semi-arid belts to even small changes in water availability can be explained in terms of Langbein and Schumm's sediment–yield curves and the role of changing vegetation cover, though the effects of both rainfall and temperature changes are difficult to disentangle.

There is, however, growing evidence, much of it still ambiguous, that considerable climatic changes affected even the interior of the deserts, and possibly other climatic zones as well. Pollen analysis in the Sahara has shown that both Mediterranean and Guinean floral elements invaded the interior deserts in the late Pleistocene, and similar work in East Africa and Angola had suggested vertical shifts of the vegetation belts of several hundred metres. It has been objected that we know so little of the synecology of the present vegetation that we cannot legitimately infer climatic changes from scattered pollen records, but many lines of evidence taken together lead to a strong presumption of major changes, perhaps with considerable regional variability. What happened to the rain-forests in the Pleistocene is one of the yet unsolved problems of climatic geomorphology. It is often argued that because of their species composition and diversity the tropical forests must have undergone only marginal fluctuations at their latitudinal and altitudinal boundaries. The appearance of the present forest may be a poor indicator of antiquity, however. Charcoal layers under supposedly undisturbed Nigerian forests, archaeological remains under Central American and South-East Asian forests, demonstrates the rapidity of regeneration after disturbance. Relict desert sands in Central Africa extend as far north under rain-forest as the mouth of the Congo, though now restricted to the Kalahari.

Quaternary climatic changes were essentially short-term changes of great intensity. Though the length of the Quaternary is itself in doubt, there is now some agreement on a figure of 2 million years, rather than the much shorter figures previously accepted. 'Pre-Quaternary' glaciations now being described from Iceland and elsewhere are dated at 5 million years, and may make an even longer Quaternary necessary. During this time there were at least four, and probably more, major climatic fluctuations, apparently synchronous in high latitudes all over the globe, and a very large number of minor fluctuations both in glacials and interglacials. If it is accepted that particular climate–vegetation conditions generate specific sets of landforms, then the effect of these Quaternary changes would depend on the rate at which landforms adjust to new equilibrium conditions. Such rates will be a function of not only the magnitude and duration of the processes involved, but of the nature of the pre-existing landforms and the resistance of the rocks. Except in the case of ice action, it is likely that many climatic conditions existed for so short a time that they were morphologically significant only in areas of weak rocks and considerable relief.

In a sense the Quaternary has been so complex that the greater part of it has been ignored, especially as later changes tended to obliterate the effects of earlier ones. Much effort has gone into reconstructing the climates and landforms of the Holocene and of Wisconsin times, and only recently has comparable attention been given to the geography of the last interglacial. Little is known about earlier times, covering the great part of the Quaternary. There is the added complication that this period was one of major tectonic movement, with rifting and vulcanicity in East Africa, and uplift of the Andes in South America and comparable mountains in South-East Asia.

In spite of their intensity, therefore, it is possible that Quaternary climatic changes were less important as landscape-forming agents, because of their short duration, than earlier climatic conditions. The cooling of the Tertiary in northwest Europe has long been known on geological grounds, and warmer sea temperatures at the end of the Mesozoic have been deduced from isotopic measurements on fossil foraminifera. Baulig and many others, particularly in southern Europe, have described deep weathering profiles, laterite-like deposits, and low-angle surfaces which they ascribe to humid tropical conditions during the Tertiary. Strakhov [1967] has attempted to summarize the evidence for Tertiary climatic changes on a world scale. In most areas, however, present landscapes are complex mosaics consisting of small areas inherited from Tertiary conditions, and tracts of forms developed during the Quaternary complex climatic conditions.

Taken together, the evidence suggests that climatic changes have been so continuous in the last 50 million years, and so rapid in the last 2 million years, that equilibrium landforms can rarely have been developed. Davis considered his 'climatic accidents' to be infrequent replacements of one type of climate and landscape by another, but his concept may have to be modified to take account of continuous climatic changes and the constant readjustments of denudation systems to them.

4. Conclusion

Climatic geomorphology rests on the assumptions that different climatic inputs in the denudation system will result in different landform outputs, and that equilibrium landforms will differ between climatic regions in such measurable parameters as drainage density, maximum slope angle, and slope form, as well as in the development of characteristic type-landforms. Many of these differences have yet to be demonstrated, and some at least of the supposedly characteristic forms may have diverse origins. Climate itself is an elusive quantity to define, and most previous workers have used such coarse measures as mean annual rainfall and mean annual temperature. What is required is, first, data on water availability and temperature conditions at the ground surface and in the weathered mantle, where areal movements and weathering changes take place, and second, some information on the water and sediment outputs from the system, through analysis of river discharges, hydrographs, and sediment yields.

In the absence of such data judgement must be suspended on the reality of climatic control of fluvial landforms, though clearly the arid and glacial landforms differ fundamentally from those of more humid areas. In the most intensive analysis yet made of discharge/erosion relationships in contrasting climatic areas, Douglas has concluded that climate as such is of small significance, at least as traditionally understood, and that we are dealing simply with different combinations of the same processes. Similar sediment yields could, he believes, be provided by similar rainfall seasonality conditions under what appear to be very different climatic conditions. Most of the analyses so far made of climatic–morphologic cycles have been highly deductive: and it is, of course, possible to construct internally coherent and logically consistent schemes relating climatic characteristics to weathering processes and rates, vegetation types, run-off characteristics, erosion rates, and landform development. Such schemes require more intensive testing in the field than they have yet received.

If landform development is viewed as the adjustment of form to process in denudation systems, the emphasis of climate at the expense of other factors, particularly lithology and structure, is bound to be distorting. It is possible that at certain scales of landform development climatic parameters are important independent variables, whereas at others they are dependent or not significant. The investigation of such scale-linkages in the correlation structure of diverse denudation systems is the main task facing climatic geomorphology.

REFERENCES

BECKINSALE, R. P. [1957], The nature of tropical rainfall; *Tropical Agriculture*, **34**, 76–98.

COMMON, R. [1966], Slope failure and morphogenetic regions; In Dury, G. H., Editor, *Essays in Geomorphology*, (Heinemann, London) pp. 53–81.

EYLES, R. J. [1966], Stream representation on Malayan maps; *Journal of Tropical Geography*, **22**, 1–9.

GEIGER, R. [1965], *The Climate Near the Ground* (Harvard University Press, Cambridge, Mass.), Revised edition.

LANGBEIN, W. B. and SCHUMM, S. A. [1958], Yield of sediment in relation to mean annual precipitation; *Transactions of the American Geophysical Union*, **39**, 1076–84.

PEEL, R. F. [1966], The landscape in aridity; *Transactions of the Institute of British Geographers*, **38**, 1–23.

PELTIER, L. C. [1950], The geographic cycle in periglacial regions as it is related to climatic geomorphology; *Annals of the Association of American Geographers*, **40**, 214–36.

PELTIER, L. C. [1962], Area sampling for terrain analysis; *Professional Geographer*, **14** (2), 24–8.

STODDART, D. R. [1968], Climatic geomorphology: review and reassessment; *Progress in Geography*, 1.

STRAKHOV, N. M. [1967], *Principles of lithogenesis;* Volume 1. Tomkeieff, S. I. and Hemingway, J. E., Editors, (Oliver and Boyd, Edinburgh and London).

TRICART, J. and CAILLEUX, A. [1965], *Introduction à la géomorphologie climatique;* (S.E.D.E.S., Paris).

11.II. Geomorphic Implications of Climatic Changes

S. A. SCHUMM

Department of Geology, Colorado State University

As we view our familiar environment the question arises, did it always appear so? Major climatic changes are known to have occurred during the past million years of earth history, and over vast areas of the earth evidence of ice action dominates the landscape. In this chapter we concern ourselves only with the long-term effects of climate change on the hydrologic cycle and on the landscape, while omitting treatment of the more obvious effects of glacial and periglacial climates.

Any generalization concerning changes of climate can be very much in error for a given locality, but much evidence has been compiled to suggest that during the last million years average temperatures could have ranged from 10° below to 5° F above present average temperatures. In most, although not all, regions higher average precipitation was associated with the lower temperatures of continental glaciation, and average precipitation was, at least for some presently arid and semi-arid regions of the United States, about 10 in. greater. During inter-glacial time and during a brief post-glacial episode higher temperatures prevailed, and average precipitation was about 5 in. less than that of today (Schwarzback, 1963).

The effects of climate changes of these magnitudes are not direct. Rather, with changing climate the relations between climate, runoff, and erosion are altered by significant changes of vegetation. Geomorphic evidence of a climate change is, in fact, evidence only of a change in the hydrologic variables of runoff and sediment yield. Therefore, the relations between climatic, phytologic, and hydrologic variables must be considered before the effect of a climatic change on the landscape can be evaluated.

1. Sediment movement

Modern hydrologic data from the United States have been used to demonstrate climatic influences on the quantity of runoff and sediment delivered from drainage basins. The family of curves of fig. 11.II.1 illustrate the general relation between climate and runoff (Langbein *et al.*, 1949). The curves show that, as might be expected, annual runoff increases as annual precipitation increases. However, runoff decreases as temperature increases with constant precipitation because of increased evaporation and water use by plants.

The relations between annual sediment yield and annual precipitation and temperature for drainage basins averaging about 1,500 square miles in the United States are presented in fig. 11.11.2. The 50° F curve of fig. 11.11.2 shows the relationship between sediment yield and precipitation adjusted to a mean annual temperature of 50° F (Langbein and Schumm, 1958, p. 1076). Sediment

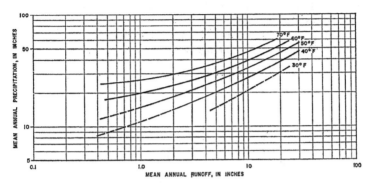

Fig. 11.11.1 Curves illustrating the effect of temperature on the relation between mean annual runoff and mean annual precipitation (After Langbein *et al.*, 1949 and Schumm, 1965).

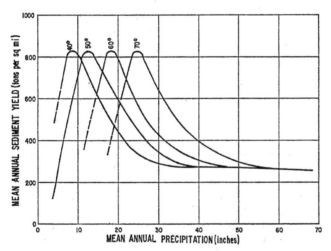

Fig. 11.11.2 Curves illustrating the effect of temperature on the relation between mean annual sediment yield and mean annual precipitation (From Schumm, 1965).

yield is a maximum at about 12 in. of precipitation, but it decreases to lower values with both lesser and greater amounts of precipitation. The variation in sediment yield with precipitation can be explained by the interaction of precipitation and vegetation on runoff and erosion. For example, as precipitation increases above zero, sediment yields increase at a rapid rate, because more runoff becomes available to move sediment. Opposing this tendency is the influence of vegetation, which increases in density as precipitation increases. At about 12 in. of precipitation on the 50° F curve the transition between desert shrubs and grass occurs. Above about 12 in. of precipitation on this curve sediment-yield rates decrease under the influence of the more effective grass and forest cover. Elsewhere in the world, where monsoonal climates prevail, sediment-yield rates may increase again above 40 in. of precipitation under the influence of highly seasonal rainfall (Fournier, 1960).

The sediment-yield curves for temperatures of 40°, 60°, and 70° F are dis-

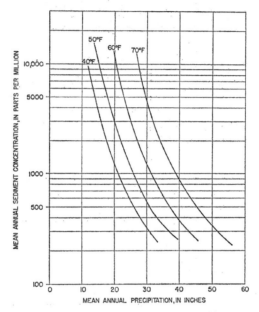

Fig. 11.11.3 Curves illustrating the effect of temperature on the relation between mean annual sediment concentration and mean annual precipitation (From Schumm, 1965).

placed laterally with respect to the 50° F curve (fig. 11.11.2). Together they indicate that, as annual temperature increases, maximum sediment yields should occur at higher amounts of annual precipitation. That is, higher annual temperatures cause higher rates of evaporation and transpiration, and less precipitation is available to support vegetation. Runoff is less, and so the maximum rate of sediment yield shifts to the right.

In addition to the amount of sediment moved, its concentration in the water by which it is moved is important. Curves were developed to show the relation between average sediment concentrations and average precipitation at different temperatures (fig. 11.11.3). For a given annual precipitation, sediment concentrations increase with annual temperature, whereas for a given annual temperature sediment concentrations decrease with an increase in annual precipitation.

One important point to be made with regard to figs. 11.11.2 and 11.11.3 is that, although more sediment is moved from a drainage system under semi-arid conditions, nevertheless sediment concentrations are greatest in arid regions. During a period of years the small number of high-concentration runoff events that occur in arid regions cannot transport the quantities of sediment that are moved by the greater number of lower concentration runoff events in semi-arid regions.

The curves of fig. 11.II.2 show that major changes in erosion rates will occur with relatively minor changes of climate if plant cover adjusts significantly to the climate change. That is, at a mean temperature of 50° F a small change of precipitation anywhere between 0 and 20 in. should elicit a significant hydrologic and geomorphic response, whereas this should not be the case in humid regions, because a major change of vegetational type and density will not accompany a small change of precipitation or temperature.

Changing type and density of vegetation exert a major control on landforms. For example, as the density of vegetation increases with annual precipitation (Langbein and Schumm, 1958, fig. 7), the rate of erosion of hill-slopes should decrease in semi-arid and sub-humid regions. However, in initially arid regions the major increase in runoff that occurs with increased precipitation (fig. 11.II.1) probably will more than compensate for any increase in vegetal cover. Therefore initially arid slopes should be subjected to more intense erosion. Although this last statement is conjectural, the peaks of the sediment–yield curves (fig. 11.II.2) demonstrate that more sediment is exported from semi-arid than from arid regions, and some of this increase must be derived from the hill-slopes.

2. Changes in the channel system

Moving down off the hill-slopes to valley floors, evidence concerning the response of river systems to climate change appears. Investigations into the relations among climate, runoff, and the character of channel systems indicate that within a given climatic region both the total length of channels per unit area (drainage density) and the number of channels per unit area (channel frequency) increases as annual runoff increases and as flood volumes increase. Therefore, increased runoff should cause lengthening of the drainage network and the addition of new tributaries to the system. This conclusion is based on data collected from within one climatic region, where, in fact, differences of soil type and geology exercise a dominant influence on both runoff and drainage density. Therefore, when the increase in runoff is accompanied by a major change in vegetational characteristics the results may be different. If a major climatic change causes a shift in vegetational type from shrubs and bunch grass to a continuous cover of grasses or from grasses to forested conditions it appears that the increased density of vegetation should prevent an accompanying increase in the length and number of drainage channels. In fact, worldwide measurements of terrain characteristics show that stream frequency is greatest in semi-arid regions, least in arid regions, and intermediate in humid regions (Peltier, 1959). It appears, then, that a major increase in precipitation can either increase or decrease the length and number of channels, and if drainage density were to be substituted for sediment yield on the ordinate of fig. 11.11.2 the curves should indicate in a very general manner the variation of drainage density with climate.

In summary, both drainage density and sediment yield should be greatest in semi-arid regions. The occurrence of maximum drainage densities and maximum sediment yields under a semi-arid climate suggests that high sediment yields are a reflection of increased channel development and a more efficiently drained system. Hence, a shift to a semi-arid climate from either an arid or a humid one should allow extension of the drainage network with increased channel and hillslope erosion.

Obviously the changing hydrology of the small upstream drainage areas and their hill-slopes will be reflected in the behaviour of the primary river channels that transport runoff and sediment to the sea. Again, the initial climate or the climate existing before a climate change is of major importance, for it determines the type of river to be considered. For example, in initially humid regions the rivers are perennial, and they remain perennial during a change to a wetter, cooler climate, although the channel will enlarge to accommodate the increased runoff.

The situation differs somewhat for an initially semi-arid region, for even the major rivers are initially either intermittent or characterized by long periods of low flow. The smaller tributaries are ephemeral, as are many in humid regions, but many more drainage channels should be present. With a shift to a wetter climate and greater runoff the flow in the major rivers and in many of the tributaries becomes perennial. Vegetation becomes denser, obliterating the smallest channels. As runoff increases, sediment yields and sediment concentrations decrease (fig. 11.11.1, 11.11.2, and 11.11.3). The result will be enlargement of the main channels. With a return to semi-arid conditions, the sediment yield increases, as both hill-slope and channel erosion increases, and runoff decreases. Deposition in and decrease in size of the main channels should result, as the tributaries again reach their maximum extent and number.

It has not always been recognized that the changes in tributary channels might not conform to changes along the larger rivers. This may not be important in humid regions, but it becomes of major importance in arid regions, where unfortunately few hydrologic data are available from which one may estimate river response to climate change. Nevertheless, increased precipitation should increase runoff in arid regions, but probably not to the extent that ephemeral rivers will be converted to perennial ones. Increased runoff should enlarge the tributary channels and extend the drainage network (arroyo cutting). Sediment will be flushed out of the tributary valleys into the main channels during local storms, and because the loss of water into the alluvium of the main channels is appreciable, aggradation of the main channels will result. In effect, increased runoff in arid regions will erode the tributary valleys, and this sediment will move downstream, where at least part of it will be deposited in the major river channels.

Without perennial flow, semi-arid and arid rivers probably are always characterized by a relative instability, and channel cutting can alternate with phases of aggradation. These events can be considered a natural part of the cycle of erosion of these ephemeral stream channels.

Deductions based on morphologic evidence have been made concerning channel adjustment to climate change, but what evidence exists that the postulated changes may be real? The evidence lies in studies of modern river behaviour in response to differing conditions of runoff and sediment yield. For example, many investigators have demonstrated that the width (w), depth (d), meander wavelength (l), and gradient (s) of rivers are related to the average quantity of water or the discharge (Qw) passing through a channel. As shown by equation (1), channel width, depth, and meander wavelength will increase with an increase in discharge, but gradient will decrease.

$$Qw \simeq \frac{w, d, l}{s} \tag{1}$$

Although the relation between these channel parameters and discharge are highly significant, nevertheless, for a given discharge a ten-fold variation in meander wavelength, width, and depth can occur. Recent work has demonstrated that with constant discharge an increase in the bedload (Qs), the quantity of sand and coarser sediment moved through a channel, will cause an increase in channel width, gradient, and meander wavelength but a decrease in channel depth, as follows:

$$Qs \simeq \frac{w, l, s}{d} \tag{2}$$

Channel shape is significantly influenced by sediment load, and as indicated by equation (2), an increase of bedload at constant discharge will cause an increase in the width–depth ratio, as a channel widens and shallows. Accompanying this change of shape is an increase in meander wavelength (l) and an increase in gradient (s). These changes are brought about by a decrease in the sinuosity, that is, a straightening of the course of the stream, which decreases the number of bends and steepens the gradient.

A glance at equations (1) and (2) reveals that changes in type of sediment load and discharge do not always reinforce one another, that is, although an increase in water discharge will increase depth, an increase in bedload will decrease it. The changes of river morphology that occur, therefore, depend on the magnitude of the changes of discharge and sediment load.

Evidence of river changes in response to climate fluctuations is difficult to obtain, because in most instances the adjustment destroys the pre-existing form, and the basis for comparison is lost. However, the Riverine Plain of New South Wales, Australia, is a unique area from which significant observations concerning river adjustment to changed climate can be obtained. Across this alluvial plain the Murrumbidgee River flows to the west, and evidence is visible on the surface of the plain of older, abandoned channels that functioned during different climatic episodes of the past (fig. 11.11.4). Oxbow lakes, which are remnants of an abandoned channel that was much larger than the modern river, are visible on the floodplain of the Murrumbidgee River. The channel shape and sinuosity of this channel is similar to that of the Murrumbidgee River, but its width, depth, and meander wavelength are larger. The abandoned channel is filled primarily with silts and clays, which must have comprised the sediment load of that channel and which is the predominant load of the modern river. Apparently, as a result of increased precipitation, much higher discharges moved out of the source area in the recent past; however, little change in the nature of the

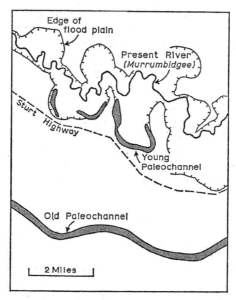

Fig. 11.11.4 Diagram made from an aerial photograph of a portion of the Riverine Plain near Darlington Point, New South Wales. The sinuous Murrumbidgee River, which is about 200 ft wide, flows to the left across the top of the figure (upper arrow). It is confined to an irregular floodplain on which a large oxbow lake (youngest paleochannel, middle arrow) is preserved. The oldest paleochannel (lower arrow) crosses the lower part of the figure.

sediment load occurred, and although the size of the channels are very different, they are in other respects morphologically similar. The contact between the Murrumbidgee River floodplain and the surface of the Riverine Plain proper (fig. 11.11.4) is a series of large meander scars, which are further evidence of the large size of this most recent palaeochannel. Another older set of abandoned channels can be detected in this area (fig. 11.11.4), and these palaeochannels are morphologically completely different from the palaeochannels, which once occupied the floodplain, and the modern river. The older palaeochannel (fig. 11.11.4) is straighter, wider, and shallower than both of the younger channels, and because of its straight course its gradient is twice that of the modern river. Its abandoned and aggraded channel is filled with sand and fine gravel, indicating that this channel was moving a very different type of sediment load from the source area, during presumably a more arid climate. The differences among these three channels can be explained by the changes in water discharge (Qw) and type of sediment load (Qs), as shown by equations (1) and (2).

Although direct evidence of the response of landforms to climate change is rare, the fact that it is the climatic effect on runoff and sediment yield that causes adjustment of river channels makes available other sources of information, as for example, downstream changes of river character with changing discharge and sediment load.

An excellent example of channel changes with changing discharge and sediment load is provided by the Kansas River system, which is tributary to the

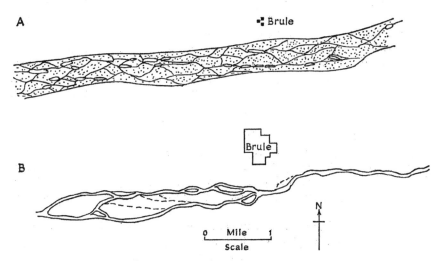

Fig. 11.11.5 South Platte River at Brule, Nebraska.
A. Sketch of channel from U.S. Geological Survey topographic map of the Ogallala Quadrangle, based on the field surveys of 1897.
B. Sketch of channel from U.S. Geological Survey topographic map of the Brule and Brule SE Quadrangles, prepared from aerial photographs taken in 1959 and field checked in 1961.

Missouri River in the western United States. Rising in eastern Colorado, the Smoky Hill River flows to the east. It drains a region of sandy sediments, and it is a relatively wide and shallow stream of steep gradient and straight course. In central Kansas two reservoirs retain much of the bedload of the Smoky Hill River, and below these reservoirs two major tributaries join the Smoky Hill River. These tributaries drain a region of fine-grained sedimentary rocks, and they introduce into the Smoky Hill River large quantities of suspended sediment (silt and clay). Although discharge is increasing in a downstream direction, the width of the channel decreases and its depth increases as a result of the influx of fine sediment and the decrease of bedload. The channel becomes more sinuous and gradient, width–depth ratio, and meander wavelength decrease. Farther downstream the Republican River, a major tributary, which drains an area about equal in size to that of the Smoky Hill River, joins the Smoky Hill to form the Kansas River. Below the junction of the two rivers, the major increase in discharge and the addition of large quantities of sand cause a major change in the channel morphology. The channel width increases markedly, as does the width–depth ratio and gradient, whereas sinuosity decreases and channel depth remains relatively unchanged.

The above changes in channel characteristics have occurred along one river as the nature of the sediment load was altered by changes in runoff and sediment yield from tributary basins. In addition, man has modified the flow patterns and sediment loads of rivers, and these changes often duplicate the effects of a climate change. Therefore, although geologic evidence is limited, the engineering literature is a fruitful source of information concerning river adjustment to man-induced changes of hydrologic variables. In fig. 11.11.5 a major change in the width of the South Platte River is illustrated. Both the North and South Platte Rivers in Nebraska were classic examples of braided rivers during the latter part of the nineteenth century, however, due to regulation of discharge, these channels have recently undergone major changes of dimensions and form. In the case of the South Platte River flood peaks have been reduced, and in response the river has changed from a wide braided channel to a narrow and somewhat more sinuous channel. Depth has probably also decreased. Similar changes have occurred along the North Platte River as a result of both a reduction in peak discharge and annual discharge. In these examples a major reduction in channel size has occurred as a result of reduced discharge, as indicated by equation (1).

Discussions of the adjustment of landforms to long-term climate changes are still highly speculative, but a more complete understanding of the possible changes that can occur is being developed. The ability to predict landscape changes resulting from a modification of the climatic or hydrologic characteristics of drainage systems has practical implications, for if man persists in his efforts to modify not only the hydrologic regime of river systems but also the climate of drainage basins, then he should be prepared to evaluate the consequences of these acts not only in terms of changed river discharge and sediment yields but also in changed channel and drainage basin morphology.

REFERENCES

FOURNIER, M. F. [1960], *Climat et erosion* (Presses Univ. France, Paris), 201 p.

LANGBEIN, W. B. *et al.* [1949], Annual runoff in the United States; *U.S. Geological Survey Circular* 52, 14 p.

LANGBEIN, W. B. and SCHUMM, S. A. [1958], Yield of sediment in relation to mean annual precipitation; *Transactions of the American Geophysical Union*, 39, 1076–84.

PELTIER, L. C. [1959], Area sampling for terrain analysis; *Professional Geographer*, 14, 24–8.

SCHUMM, S. A. [1965], Quaternary paleohydrology; In WRIGHT, H. E. and FREY, D. G., Editors, *The Quaternary of the United States* (Princeton Univ. Press), pp. 783–93.

SCHWARZBACK, M. [1963], *Climates of the Past*; (Van Nostrand Co., New York), 328 p.

Index

Milton Keynes UK
Ingram Content Group UK Ltd.
UKHW040102071024
449327UK00019B/743